PRIVACY 3.0

PRIVACY 3.0

Unlocking Our Data-Driven Future

RAHUL MATTHAN

HarperCollins *Publishers* India

First published in India in 2018 by
HarperCollins *Publishers*
Building No 10, Tower A, 4th Floor, DLF Cyber City, Phase II,
Gurugram – 122002
www.harpercollins.co.in

1 2 3 4 5 6 7 8 9 10

P-ISBN: 978-93-5277-988-8
E-ISBN: 978-93-5277-989-5

The 'Sunday Sentiments' columns on pp. 149-154. 166-167, 169-171, 172-175 and 195-197 have been reproduced with permission from the *Hindustan Times*.

Typeset in 11/15 Berling LT Std at
Manipal Digital Systems, Manipal

This book is produced from independently certified FSC® paper to ensure responsible forest management.

For Ahalya and Dhruv

Contents

PROLOGUE
The Year Aadhaar Exploded

In 2017, Aadhaar, the Government of India's grand plan to provide a unique digital identity to the 1.3 billion people living in the country, exploded.

It was fast becoming evident that not only was Aadhaar not going to remain optional as originally advertised, it was also going to become ubiquitous. The government had already made Aadhaar linkage mandatory to avail of certain schemes run through the Public Distribution System (PDS) and the LPG distribution scheme. During the course of the year many more services got added to that list. The government told regulated entities like banks and telecom companies to get all their customers to link their accounts to Aadhaar by the end of the year. As a result, customers began to be bombarded, almost on a daily basis, with messages to link their accounts to avoid disruption of services. But as much as the central government was forcing government agencies and regulated entities to ensure that their users linked their Aadhaar numbers to their accounts, the real spike in Aadhaar usage came as a result of the actions of the private sector.

In 2016, the Unique Identity Authority of India (UIDAI) had notified the Aadhaar (Authentication) Regulations, establishing a framework within which private enterprises could utilise its identity architecture for their transactions. Shortly thereafter, both the Reserve Bank of India (RBI) and the telecom regulator introduced amendments in their Know Your Customer (KYC) regulations allowing Aadhaar-enabled e-KYC to serve as a substitute for the cumbersome manual authentication process that was being followed. Relying solely on Aadhaar-based paperless SIM activation, Reliance Jio, the country's most recent entrant into the telecom sector, was able to create a world record by crossing 16 million new subscribers in its first month of operations.

Aadhaar was being used everywhere – by fleet taxi operators to on-board their drivers, by payment banks to enrol new customers and by bike sharing companies to authenticate their users. A number of futuristic business cases were being actively evaluated, including the use of Aadhaar-based paperless travel at airports, where boarding passes would be completely done away with and all we would have to do was show our fingerprints at all the checkpoints in the airport, all the way to the boarding gate and on to the aircraft.

But even as enthusiasm for Aadhaar was growing, opposition to it rose to a crescendo. Activists, lawyers and politicians of every hue demanded the cancellation of the project, calling into question the technology, the fact that it had been implemented without legislative backing, and that rather than providing benefits to the poor it was resulting in exclusion. They claimed that the sinister motive behind linking anything and everything to the Aadhaar database was the creation of a giant panopticon to watch over us and monitor our every move. There was a real fear that we had already slipped into an Orwellian dystopia where the State has a record of everything we do, where we go and who we meet.

An increasing number of instances were coming to light where the flaws in the technology were being exposed. In one instance, a company that was supposed to provide enrolment services developed a programme that used a stored biometric to conduct multiple transactions without the need for human intervention.[1] In another, e-KYC information from the UIDAI database was made available allegedly by exploiting an HTTPS weakness.[2] Both these instances turned out to be far more innocuous than originally reported but they tapped into the rising consternation around the vulnerability of one of the largest government databases of personal identity on the planet. There were rumours that the foreign companies that provided the biometric algorithms that were used in the enrolment process were leaking personal information out of the country to the foreign governments under whose jurisdiction they were based, giving rise to an unreasonable fear of an invisible foreign hand.

Never before had a single technology so radically polarised the nation. On the one hand, there were those who were excited by the many opportunities that had opened up, now that they had access to a digital identity that was capable of frictionless interaction across multiple services. On the other, the doomsayers only saw all the harms that would arise if this powerful technology was misused. These two extreme views controlled much of the narrative around Aadhaar but neither of them was entirely correct. Done right, Aadhaar had the potential to be the defining technology of this generation. Done wrong, it could set the country back decades.

Opposition against Aadhaar was largely focused on the privacy implications of the project. It might have been a bit more muted if we actually had in place a full-fledged privacy law. Or if somewhere in our judicial history we had recognised personal privacy as needing protection under the law. Not only were we among the very few countries in the world that did not have a privacy law, a few years previously, while defending the Aadhaar project, the attorney general of the country had argued that India had no such thing as a fundamental right to privacy.

How did we get here? What twist of fate, what curious combination of unfortunate circumstances led the largest democracy in the world to function for seven decades without a privacy law – to the point where the highest legal officer of its government refused to even acknowledge its existence as a basic and natural human right?

As much as this is not a book about Aadhaar, it would not have come to be written but for it. As a technology lawyer, I have always had more than a passing interest in privacy and its implications on technology. But my interest has been limited to interpreting the law as it applies to my clients. Aadhaar made me examine how changing technology influences our notions of privacy and called into question many of my long-held beliefs about how data must be regulated.

We are so obsessed with the harms that can befall us if our privacy is violated that we have designed our laws to prevent that from occurring at all costs. As a result, our privacy laws are restrictive, designed to prevent us from doing anything that could even accidentally result in such damage. It has made us overly cautious – to the point where we have begun to miss out on the benefits that technology can offer. But technology rarely heeds to legal restrictions. Innovation has always proceeded apace despite the attempts of the law to restrain it. Every time technology found a new way to inveigle itself into our personal space, we have so enjoyed the benefits that it brought that, rather than prohibit it, we have adapted our privacy regulations to account for it.

Aadhaar made me wonder whether we are, once again, at one of those crossroads in the evolution of privacy jurisprudence where there is a need to re-examine the laws that currently govern us to see if there might be another, smarter way to protect our personal privacy.

To do this I felt I needed to better understand the origins of our current notions of privacy – to see if there is something in its past that will inform its future. This book is the product of that research. It attempts to provide some explanation for how we have come to our current notions of personal space and individual privacy, starting from early human tribes in whose egalitarian social structures privacy was all but non-existent, all the way down to our data-driven present where it seems there is little we can do to conceal our thoughts and actions from those around us.

In the process of researching this book, I came to realise that privacy has passed through three distinct phases of evolution. In the first stage of privacy we developed, for the first time, the idea of personal space and private thoughts. We evolved laws that were designed to protect information that we disclosed to others in confidence, allowing us recourse against those who betrayed our trust. But as technology developed, this approach was not nearly sufficient to protect our personal privacy. We needed protection from those beyond our immediate circle of trust as technology made it possible for complete strangers to invade our personal space.

Privacy 2.0 was about elevating it to the status of a right that could be exercised against anyone who impinged upon our privacy without our express consent. This is the construct upon which most of our laws have been based and on which we have relied for years to protect us. It has served us well from the time when data about us was kept in physical files all the way to the dawn of the internet when everything began to be digitised.

We stand at the threshold of Privacy 3.0. Our world today is so rich with data that the consent-based protection that has served us well for so many decades is proving ineffective against the onslaught of modern technologies. There is a need to re-imagine privacy law to allow us to function in the modern data-driven world that we find ourselves in. And I have an idea about how we might go about that.

PRIVACY 1.0

1
Naturally Private?

It may be the most perfect photograph I have ever taken. The pair of zebras are well positioned in the foreground, their bodies lined up with each other so symmetrically that, were it not quite obviously the middle of the African grasslands, one might have been forgiven for thinking they were posing. Each zebra had its head on the other's neck, almost as if they were snuggling in a close embrace. As if to prove that it is always possible to improve on a good thing, an oxpecker gently fluttered down, perching itself on the butt of one of the zebras.

Just after I took the shot, there was a titter of excitement in the safari van. Off to the left a lioness had appeared, her eyes and head locked in one position, her body absolutely still and focussed ahead of her on the zebra herd. As I watched, her head dipped and her shoulders bunched up as she slipped into the characteristic stalking position. She was crouched so low that her belly scraped the ground, consciously reducing her profile so that her entire body was completely hidden by the small patch of long grass separating her from the zebras. She had identified the herd and selected a target. We were about to witness a kill.

I looked back at my zebra couple, hoping it was not them she was after. They had seemed so lost in each other that the lioness could have stepped out in front of them and they would not have known it. To the contrary, when I looked at them I noticed an electricity that was previously not there. One of them had turned her head around in the direction of the lioness and the other was almost quivering with the effort of straining to identify the danger that they had instinctively sensed. They were both keenly aware that something was amiss but not yet sure what it was, and were unwilling to raise an alarm without proof that there was something to be alarmed about.

The lioness must have twitched, or the wind moved the tall grass enough to give one of the zebras a glimpse of a tawny back or tail. One of them let out the unmistakeable two-snort sound that every creature in Masai Mara knows is the alarm call of the zebra. Instantly, the entire herd, almost as one, dropped whatever they were doing and galloped away. Within minutes the Mara was deserted again and the lioness was left staring at their dwindling backs.

She straightened up in disgust and walked away.

Zebras are among the most striking creatures in Masai Mara. Their unique striping patterns help them stand out against the background, making them ideal subjects for the wildlife photographer. But for all their monochrome beauty, they are among the most vulnerable creatures on the plains. They are grass-eaters, who, unlike the bigger herbivores they share the Mara with – the hippopotamus and the elephant – don't have great big scimitars for teeth to protect them when they are attacked. The zebra's only defence against death is fleetness of foot. If they can get enough of a head start, they have the acceleration and stamina to be able to escape from almost any one of their natural predators. If they can't, the sheer strength and power of the big cats will bring them down.

My photograph of the two zebras is one of the most widely appreciated images I have taken in over two decades of amateur wildlife photography. I titled it 'Zebra Valentine', alluding to the almost human affection that these animals seemed to be displaying, lost in their own company in the midst of all the untamed wildness. But there is much this photograph doesn't say. This wasn't some canoodling couple I had come across. What I had captured was two sentries standing in the ideal defensive position, each one covering the blind spot of the other, straining every sinew to spot danger before it spotted them. But for this defensive construct, the herd might not have been able to spot the crouching lioness and the danger she represented. Romance couldn't have been further from their minds. All they were thinking about was survival.

This is why zebras move about in herds. By throwing in their lot with other zebras, they spread risk across the larger group, allowing weaker animals to leverage the strength of numbers to improve their odds of survival. As the English geneticist, Francis Galton, once observed, the primary purpose of animal herding is to reduce their vulnerability to surprise attacks. By committing to a communal existence, individual members of the herd get to be a 'fibre in a vast sentient web',[1] significantly multiplying their individual faculties. Herding allows individuals to forage, nurture their young, and sleep without worrying about predators, reassured that others in the group are keeping a watch for them. Since they operate in such close proximity with each other, they become 'the possessor of faculties always awake, of eyes that see in all directions, of ears and nostrils that explore a broad belt of air'.[2] This transforms the herd into a communal sensory organism able to simultaneously observe the world, forage for food and rest. It may choose to place greater emphasis on defending the young and child-bearing females so that the herd can improve its chances of propagating the next generation.

It is impossible for any one animal to always be on the alert. They need to forage and rest, and at any given time they are susceptible to predation. On the other hand, moving about in groups greatly enhances their chances of survival. As a result, herding is one of nature's most widely used survival techniques. All over the planet – on land, in the air and in the seas – living creatures band together to take advantage of proximity. With the combined strength of a large enough group, even the smallest animals are able to stand down their fiercest natural predators.

Solitude is a luxury enjoyed by only those few animals who occupy a position on the top of the food chain. Everyone else needs to form alliances with other members of their species so they can live together and get to live a little longer. As a concept, solitude and, consequently, privacy is utterly alien to nature, so how did human beings evolve and cleave to it so strongly that they now universally accept it as being so fundamental to the very notion of humanity as to be above the written law?

2

In the Fish Bowl

Early humans were no different from the animals they hunted. They too displayed herd behaviour, hunting in packs so that they could make up with guile and sheer numbers what they lacked in speed and brute strength. They carried this herd mentality beyond hunting, banding together into tribes to take advantage of the safety that comes from numbers, and ensuring that children and weaker members of the tribe were protected and nourished by the stronger. They delegated responsibilities – some members of the tribe assumed guard duty while the rest slept peacefully without worrying about being attacked in their sleep.

Solitude and privacy were not just unacceptable among early human societies; they were downright dangerous. Humans had nothing but their wits to use as weapons, and wits worked better in a group. Anyone going off on their own was forsaking the benefits of the group – and in those prehistoric times, that could have cost them their lives.

For that reason, early humans lived together in close proximity. They had no personal space and knew everything there was to know about each other – their likes and dislikes, strengths and weaknesses. Since the safety and well-being of the entire tribe depended on every single member pulling his weight, it was essential to the survival of the group that they knew everything there was to know about each member of the tribe. Keeping secrets was dangerous, as in an emergency hidden knowledge could mean death.

Early human tribes never stayed in one location for long. Their lifestyle depended on their being able to hunt down animals, and their source of food was the great herds that roamed the grasslands. And so they followed these great herds as they migrated from place to place, living a symbiotic existence with them – killing whenever they need to for sustenance and never letting the herds get too far away from them for fear of losing their source of food.

Over time the great herds began to dissipate, ranging further afield and dividing themselves into smaller and smaller herds, making them harder to hunt down. At the same time, tribal populations had increased to a point where the demands that were being placed on hunters grew beyond their ability to service. Hunting raids became less and less productive, as an increase in the number of mouths to feed coupled with a dwindling population of animals to hunt took an irrevocable toll on the hunter-gatherer lifestyle.

Over the years, a number of animals – pigs, goats and cattle – had opted to exchange the uncertainty of life in the pre-historic wilderness for the security of travelling in the company of humans. They followed human tribes around, feeding off the scraps that were left behind, offering themselves up to be domesticated and roaming freely about within the encampments. With the hunt becoming a less than reliable way to find food, man began to turn to these animals for sustenance. They soon realised that it took far less effort to use animals like goats and chickens for food compared to what they risked trying to hunt animals in the wild.

As global temperatures increased and the weather grew moister and more humid, cereals and legumes began to proliferate, offering humans new nutritional possibilities. Man began to figure out how to modify his natural environment, discovering that seeds and shoots could be artificially cultivated, allowing them to make all sorts of nourishing food accessible in their backyard. With that evolved agriculture – the first technology that truly separated man from the rest of nature, making humankind the first species that could bend nature to its will. Babies began to get more grain in their diet, allowing them to be weaned sooner. This meant that women could bear more children during their lifetime than was previously possible. Since tribes were becoming less nomadic, children no more needed to be carried around from place to place and human infants began to live longer. This resulted in an overall increase in the absolute number of infants surviving to adulthood, causing a population explosion of sorts.

Humans had no option but to settle down, building permanent camps and developing rudimentary social structures to better cater to the requirements of sustaining their expanding numbers. Now that they had struck roots, no longer needing to travel around, one would have thought that it was at this juncture – during the transition from nomadic hunter-gatherers to farmers – that the concept of privacy entered human consciousness. As it happens, we have evidence that this was not the case.

Ordinarily, it would have been impossible to know how these early human societies functioned. Anthropologists usually only have excavated remains of ancient societies to go by – and there is only so much that one can discover from a shard of pottery about how life must have been in those ancient times. However, we are fortunate in that up until relatively recently there were a number of tribal societies that lived so far removed from human habitation that the very anthropologists who were planning to study them were their first contact with modern civilisation. From all accounts, when they were 'discovered', they were living in much the same way as their forebears had for millennia. By studying these primitive societies, their social dynamics and community behaviour, we could get a glimpse of what the social norms of those cultures were, and use this information to test our hypothesis about privacy in the social construct of ancient man.

From the evidence, it appears that there was, in fact, no change in behaviour once mankind settled down in permanent locations. The early villages of man had much the same social norms as the hunter-gatherers who went before them, and used the same mutual surveillance model to ensure that everyone behaved as they were expected to. Safety was still a concern. Now that they had permanent homes, their enemies and predators knew where to go to find them. The entire tribe still depended on every single member pulling his weight, and it was more dangerous than ever to keep secrets.

One such ancient society that survived into modern times was the Kalahari Bushmen – the Kung tribesmen – who were made famous by the movie *The Gods Must Be Crazy*. They are one of the oldest ethnic groups in the world. For years they lived in southern Africa on the fringes of the Kalahari Desert in a sort of hunter-gatherer, quasi-nomadic existence not very different from that of the early Stone Age man.

Kung huts were not designed to be lived in. They were places where belongings were stored to protect them from the elements, but no one ever used them to retire into. In fact, since the Kung rarely ever spent time alone, their huts weren't even designed to be habitable. Seeking solitude was regarded, among the Kung, as bizarre behaviour, and their huts were spaced so close to each other that people from one house could hand utensils to someone in another house without getting up from their hearth. We could probably go so far as to say that the layout of the Kung camp was designed to actively discourage privacy.

The Mehinacu of central Brazil took things a step further. When botanist Thomas Gregor visited them in 1967, surveillance was a way of life. Everyone was constantly watching everyone else, always peering out through openings in their dwelling units to keep an eye on what was going on. Curiosity was a virtue, and they made a concerted effort to track the activities of any tribesmen who were not physically present among them, recognising them from the imprints their heels left on the sand or the mark of their backsides when they squatted to chat. As a result of this aggressive, unrelenting surveillance, everyone knew exactly where every other member was at any given time. Their huts offered no privacy whatsoever. Everyone knew that their conversations were being overheard and that anything they said would soon be common knowledge throughout the village.[1]

Thus, even after they settled down, humans were unable to shake off the habits that had been ingrained in them by nature. They still actively abhorred privacy and lived their lives in the full gaze of everyone in the tribe. They banded together to take advantage of the safety that came from numbers. They understood that the bargain for this sort of protection was that they would have to be open and not hide anything so that everyone in the tribe could act together in each other's best interests.

The social construct of these early societies was egalitarian – everyone remained happy so long as no one tried to get ahead. Anyone who tried to do better for himself did so at the expense of someone else. If someone didn't share the food they had gathered, someone else in the tribe would have to go hungry. In the egalitarian way of life, this was unacceptable. Everyone had to be content with their lot, working for the common good so that they could all benefit. In order to be able to achieve this without a government or police force enforcing these unspoken rules, every member of the tribe had to keep an eye on everyone else, ensuring, through constant mutual surveillance, that everyone behaved. For a system like this to work, no one could have any personal space.

As a consequence, anything that blocked the flow of personal information between members of the tribe came in the way of security and was seen as prejudicial to the well-being of the other members. Anyone who actively sought seclusion from his neighbours was thought to be engaging in anti-social behaviour and was branded as untrustworthy. These societies had no concept of privacy. One might even go so far as to say that they could only survive if they ensured the complete absence of privacy.

Today this sort of behaviour seems deeply intrusive. Most of us would cringe at the social consequences of constant surveillance. But life in these egalitarian societies was not without its advantages. Since everyone knew everyone else, it was impossible for anyone to misappropriate someone else's rightful share of food by impersonation. Cheating and fraud were unheard of as everyone knew who you were, what you were good at and all your faults and flaws. You couldn't gain advantage or favour by presenting yourself as something other than what you were.

Nearly all the anthropological studies of early human tribes describe similar behaviour. Early societies not only had no concept of privacy, they were, on the other hand, organised to ensure the free flow of personal information between members of the tribe. The fact that humans were living in fixed dwelling units did nothing to change their basic attitude to privacy. The Semai of central Malaya designed their huts such that everyone in the tribe knew what was going on with everyone else. The walls were deliberately designed to be thin and the house itself kept permanently open to the breeze. Unlike the Kung huts, Semai houses were intended to be lived in – designed to protect people from wild animals and the elements, but not to shield them from each other. So much so that, among the Semai, refusing someone admission into your home was an act of extreme hostility.[2]

In Samoa, houses had no walls. Residents lived in full public view, lowering their blinds only when it rained to keep themselves dry. Everyone expected to know, at all times, what was going on inside each house. If any house was kept closed for extended periods of time, other tribespeople would begin to get suspicious about what was going on behind closed doors. In fact, walled houses were called 'palagis' – the structure that arouses suspicion – and it was presumed that people you cannot see do bad things.

However, in time, as humans began to build more permanent encampments, they had to deal with the increasing complexity of catering to their larger societies. As their villages grew in size, the effort of constant surveillance began to take its toll. Since everyone knew they were being watched, they were careful about what they did at all times. Any misstep was noticed and instantly communicated to everyone in town. And there was no let-up in the surveillance – every single moment of the day and night, everything they did had to be in conformity with what was expected of them and there was absolutely no opportunity to relax and be themselves. As a result, everyone was constantly under pressure to be on their best behaviour, knowing that any slip would result in ridicule or, worse, castigation and excommunication.

The tension began to tell and conflicts among members of the tribe began to grow. Unlike in their nomadic past, when irreconcilable differences between individuals could be resolved by one party simply walking away, now that they had invested in permanent residences, it was much harder for them to vote with their feet.

No one knows exactly when it happened, but there was a point when an innocuous development gradually crept into human life, offering residents welcome respite from the pressure of constant surveillance. It was the first time that technology affected personal privacy, but it would not be the last. From that point onwards, technology would repeatedly influence our understanding of the notion of privacy, forcing us to re-examine our perspectives time and again, re-drawing boundaries that we had previously set to allow us to deal with new challenges that these technologies bring.

The technology that started this all did something quite remarkable. It created, for the first time, the concept of self – an idea that had not truly existed till that time. In doing so, it gave birth to the individual as an entity truly distinct from the community and allowed every human being to have a personality that was unique. And which could, for the first time, be hidden from those around him.

3

What Walls Did

Humans have built shelters for aeons. These constructions have been designed to guard against the elements, to serve as a place to store possessions and, when it rained or the weather was otherwise inclement, to take shelter for a spell. Nothing in the original design or function of these shelters gave any indication that they would become the personal spaces that we think of today.

Humans started out by taking shelter in caves and under overhangs of rock. In time they learned to build shelters and dwelling units. They knew how to build walls, but in almost every instance the walls were designed to support the roof or to provide shelter from the rain. They were built using wattle-and-daub techniques that rendered them porous to the sights and sounds of the external world. Living inside these houses offered no privacy whatsoever. Conversations were audible through these walls, so, even while they were inside, no one had the remotest expectation of privacy.

When they retired at night there were no separate rooms in which they slept. Everyone – domestic animals included – lay down together on the floor in the same cramped physical space. These houses were dark and dank and, because animals and humans were all crammed together in that confined space, smelled foul. Little wonder that most families preferred to spend all their time outside their homes, going in only when absolutely needed.

When man developed technology that allowed him to construct substantial walls, he did so originally in order to build a big protective barrier around his town to improve its defensive capabilities. Ancient cities like Jericho near the Jordan River and the Sumerian city of Uruk[1] in modern-day Iraq were world famous for the impregnability of their walls.

Eventually, the technology that was used to build these city walls were extended to the construction of houses. As homes began to be built with substantial walls, humans realised that these thicker structures allowed them to insulate themselves from the sights and sounds of the world around them and, at the same time, prevented those outside the dwelling unit from determining what was going on inside. For the first time, man had a space into which he could retire and cast off the social expectation of constant vigilance.

This opportunity to escape from the constraints of conformity gave rise to a freedom of thought and action that had previously never existed. Within the private spaces of their homes, people were able to open themselves up to new ways of thinking and behaving, discarding the facade they had had to maintain when they were always watched. Walls allowed man to experience a form of mental relaxation that had never, to this point, been experienced in nature. It was, by far, one of the most significant contributions that any technology had made to the evolution of human society.

Walls facilitated a novel experience – one that encouraged honest personal self-expression in a way that humans had never previously experienced. They allowed mankind, for the first time, to savour the benefits of privacy – to create a zone within which people could shape the development of their personalities and their personal relationships differently from what was previously possible.

Once this happened, people were able to develop two distinct personas – one for public consumption that they displayed when they were out and about in the world, and the other that they only allowed to surface within the confines of their home when they were in the company of those they trusted. Over time, this 'public self' grew into an exaggerated caricature of themselves, designed solely to conform to the expectations of society. The private quarters of the home became like the green room in a theatre where man carefully layered his public image on to his private self, transforming into the image of what society wanted him to be.

We spend considerable effort today cultivating this public image using clothing, grooming and manners to portray a personality that fits the expectations of social status and public standing while suppressing our inner character and personality. Our social interactions have become dependent, almost exclusively, on the public perception of who we are. We are known by the carefully crafted personality that our business associates and social acquaintances interact with, and it is on this charade that our social standing is based. Our private self is very different, motivated by desires and passions that we strive to keep hidden lest they mar the image we have worked so hard to create. So much so that, in many instances, that private personality is so different from the public persona the rest of the world sees that it seems like a different person entirely.

It is in the evolution of this split personality that our modern notion of privacy is rooted. As long as man lived his entire life in the public gaze, there was no need for privacy. Everything there was to know about him was out there in the open for everyone to see. But the moment he could be two people at once, he needed to ensure that his private personality was always kept concealed from public view. For this reason it became increasingly important for man to find ways to protect his private image from coming into the public gaze.

Thus it was that what man's peers and the world at large thought of him came to affect his place in society and he did everything he could to improve his social perception. In ancient Greece, reputation was the primary driver of social interaction, and personal standing was the fundamental axis of urban life. In those crowded towns, a neighbour's knowledge of one's household affairs could be socially devastating. It was this politics of reputation and the concern that activities inside the house could have repercussions on their standing in public that drove architectural choices. Archaeological studies of the ruins of ancient Greek houses indicate that they were designed to be inward-facing – the ground-floor windows on the street side were situated high up on the wall, restricting the ability of those on the street to look inside.[2] The entrances to houses were narrow and angled so that the only view from the street when the door was opened was that of a blank wall in the vestibule.

That said, privacy in ancient Greece (as in almost every other civilised society since) was clearly the prerogative of the wealthy. The richer you were the more likely it was that you were going to be able to afford a large house that could contain buffers that provided greater seclusion and isolation from the public. The ancient Greeks used space to create zones that were spatially discrete, separate and out of sight of others.

Among the Romans, who succeeded the Greeks as the next great European civilisation, there was an entirely different approach to privacy. The Romans believed in flaunting their success with garish and ostentatious displays of personal wealth. Among the upper echelons of Roman society, people lived their lives as openly as possible so that the whole world could know how well-off they were. As a result, their homes were constructed with a view to make a big show of their riches. They built huge gardens open to the public, from where anyone and everyone could see what was going on inside, even the most intimate activities. For the Romans, wealth was meant to be displayed, and if in the process personal and family matters were made visible to the public, they had no qualms about letting it all play out in the open.

Even in these societies, privacy was not a luxury that was available to the poor. Everyone but the very wealthy lived in crowded urban ghettoes – small cramped spaces where no one had the choice to determine what should be kept private and what should not. However, unlike in egalitarian societies where the concept of privacy itself was unknown, in these urban villages, everyone had an understanding of what privacy was – even if not everyone could afford it.

As a result, these urban societies still displayed many of the behaviours that were part of egalitarian societies. People kept a watch on their neighbours and knew who among them was up to no good. They could tell immediately when things were amiss and were particularly suspicious of anyone who was moving about furtively and acting secretive. Even in these early urban ghettoes, there existed a fundamental mistrust of secrets and those who kept their personal affairs private.

Things might have remained that way in Europe had it not been for the influence of religion. According to the Bible, not only was it wrong to commit an evil deed, it was wrong to even think of committing one. This was a tremendously high standard to meet. The only way anyone could hope to be completely pure of thought was to abstract himself or herself from society. As a result, early Christian monks who were actively in search of spiritual achievements believed that seclusion was the means to achieve their goal. They took to living alone so that their minds could be sheltered from the distractions that gave rise to evil thoughts.

In time, these concepts of seclusion and privacy began to permeate all Western religious practices. Priests started to endorse the act of staying away from society as a form of meditation and reflection. The Church encouraged the idea that man should communicate privately with God in order to seek forgiveness for his sins. They invented the mechanism of the confession as a means to ensure the expiation of all sins (including evil thoughts).

In 1215, the Fourth Council of the Lateran declared that confessions were mandatory and applied them to all. From that point onwards, everyone was required to confess their sins, in the privacy of the confessional, to a priest who would intercede on their behalf with their maker to grant them the forgiveness they required for spiritual salvation. This practice subtly reinforced the notion that there were certain things in life that had to be kept confidential from everyone, including family members.

Thus, the Church irrevocably shifted the responsibility of moral governance from the community to the individual, making it each person's duty to police his own morality if he wanted to secure eternal salvation. Everyone was encouraged to keep their innermost thoughts secret, only sharing that information with the priesthood who could grant them salvation. The clergy were themselves under a spiritual obligation to keep confessions a secret, creating a self-perpetuating culture of confidences and concealment.

With that, the mechanism of community surveillance that had maintained social order from the dawn of human civilisation finally lost its relevance. Remnants of this culture still exist in the small sub-communities that we occupy. While there is an intangible comfort in having our neighbours look out for us when we are away, letting us know if something suspicious happens in our apartment when we are not home, there is still a line we like to draw before they get too nosy.

The open societies we lived in, in which everyone knew what everyone else did, were destroyed by the invention of substantial walls. This allowed man to develop schisms in his mind which we now identify with our individual personalities. In Europe, once the authority of the Church and the power of religious beliefs decreed that secrecy was a virtue and the only road to salvation, whatever stigma was previously attached to those who did things secretively was irrevocably washed away.

And with that our modern notions of privacy were born.

4

A Creature of Technology

Despite what policy-makers, jurists and judges will have you think, nature does not encourage privacy. On the contrary, the fact that human beings cherish their personal privacy is so out of step with nature that it is one of the features that distinguishes humankind from the rest of the animal kingdom.

Privacy was created by man. This zone of privacy is the place from which art and science, as we know them, developed. It was the basis of our cultural norms and the philosophical thoughts and values that define us. It allowed us to unlock our creativity, freeing us up to create works of art without fear of judgment or ridicule. It gave us the space to question everything around us and to find explanations for phenomena that early man could only explain as acts of God. In the process, it helped mankind develop a scientific temper, leading to new discoveries and the development of modern science and technology.

But just as privacy was born out of technology, in a curiously circular way, technology is also its darkest enemy. Every time we developed new technologies that allowed us to communicate ideas with each other in ways that were not previously possible, there were those who found ways to use these technologies to violate our confidences and make public those things that we had until then thought we could keep secret. Technology has, time and again, butted heads with privacy, forcing us to constantly re-draw the boundaries of what we had previously believed to be our personal space.

The history of privacy is the story of this tussle. Technology has played a significant role in bringing us comforts and new benefits, but, at the same time, its many unintended consequences, particularly as they relate to matters of personal privacy, constantly make us question whether those benefits are at all worth it.

It is impossible to study the evolution of modern privacy law without an understanding of the various technologies that contributed to its development. The printing press created new ways in which ideas could be disseminated, but at the same time made it possible to expose private writings and personal correspondence in ways that would have been impossible until then. The democratisation of photography resulted in the rise of the press, but at the same time encouraged the growth of yellow journalism that forced us to question where exactly we needed to draw the line between personal and public life. Post and telegraph networks that allowed ideas to travel over long distances relatively quickly shortened vast distances between people but sent everyone's personal messages over pipes that could easily be tapped into from centralised locations.

The reaction of society every time a new technology has challenged previously established boundaries of personal privacy is so consistent that it is predictable. Every time we realised that the technology we had so far thought was useful actually had harmful consequences for personal privacy, all those who were affected – usually those in the higher strata of society who had the most to lose – gathered together to protest its proliferation. In every instance, despite their protestations, the technology survived and society taught itself to adjust to account for the new challenges that these technologies posed. No technology has ever been shut down because of the privacy threat that it posed to the existing social order.

Any study of the evolution of privacy law must take place in the context of the technology that brought about the change. Our understanding of the way in which we have as a society responded to the pressures of new technologies will help inform our current anxieties around the technologies that seem to be threatening our personal privacy.

5
Confidences

Once religion gave substance to the understanding that it was perfectly acceptable to keep confidences from one another, the principles of confidentiality took firm root in our dealings with each other. Various professions borrowed these notions to artificially construct an atmosphere of confidentiality within which they could deliver their services more usefully and efficiently. Doctors assured their patients of confidentiality to encourage them to be truly honest with them about their symptoms. This allowed them to glean information about ailments that they might otherwise have been too embarrassed to disclose.

Similarly, attorney-client privilege allows lawyers to refuse to divulge information that their clients provided them during the course of legal representation – even to law-enforcement agencies that need that information to solve a crime. Similar privileges were extended, in English law, to spouses, allowing them to refrain from testifying against each other to protect the confidentiality of the marital relationship.

These principles began to insinuate themselves into the law, and eventually common law began to recognise a duty of non-disclosure in the context of fiduciary relationships and commercial contract. As best as we can ascertain, the first mention of private rights was made by English jurist William Blackstone in his *Commentaries on the Laws of England* (1765). The law, he said, has 'so particular and tender a regard to the immunity of a man's house that it stiles it his castle, and will never suffer it to be violated with impunity'.[1]

Hidden in that archaic sentence is the phrase – a man's home is his castle – often used to describe the concept of domestic privacy. It was first mentioned in 1604, in Semayne's Case, when Sir Edward Coke said: 'That the house of every one is to him as his castle and fortress, as well for his defence against injury and violence, as for his repose…'[2] and has been used many times since to describe that unique personal protection that we all are entitled to within the safety of our homes. But the idea behind the phrase goes much further than the literal words it is made up of. It uses the home as a metaphor for a private place and describes the manner in which law creates walls of regulation designed to protect against incursions into that private sphere. It describes the legal barriers that are intended to operate even when we have no walls to protect us, extending its application to invasions of privacy that occur in our social circumstance.

This metaphor – of the home as a castle – is perhaps an unintentional reference to the influence that walls have had on the notion of privacy. Even to this day, this phrase is used, freely and frequently, in debates about privacy, often without any sort of understanding of the deeper significance of the origins of the phrase. Every time it is used against me, I have to stifle a grin.

As much as the concept of confidentiality was understood and used, privacy took a long time to find its way into the law as a stand-alone concept. While a breach of confidence could be pursued and prosecuted under the law of contract or even tort, the particular value of things that are private and personal to an individual had no mention in the law. It would take the greatest invention of the time – the printing press – to demonstrate just how a technology that was capable of doing so much good could be subverted to evil. Even so, none of this might have given rise to a privacy jurisprudence had the main protagonists in two of the earliest cases not been literary royalty on the one hand and the actual English royal family on the other.

Edmund Curll was the sort of opportunistic businessman who would not have been out of place in today's world. He was constantly on the lookout for opportunities to make a profit and was not particularly constrained by scruples. He had an instinctive belief in the tremendous promise of the nascent printing industry and was convinced that the real audience for books was the common man – tradesmen, apprentices and servants – who were just not being served by the publishers of the day. With that in mind, he began to exclusively target that market, publishing cheap pamphlets on inexpensive paper, each priced at no more than a few shillings to keep them affordable. His authors were commissioned to write stories about quackery and eroticism, aimed directly at the interests of his target audience, offering them scandalous entertainment at an affordable price. And so he became the world's first purveyor of the trashy paperback.

His strategy proved to be wildly successful. Emboldened by his success, he began to publish books that were loosely based on the publications of other authors – without their consent. At the time, the world was only just coming to grips with this new printing technology. The world's first copyright law – the Statute of Anne – had just been passed and, in the early 1700s, intellectual property as a concept was at its infancy. Even though this new right created, for the first time, an artificial form of property ownership that conferred a right of ownership onto insubstantial concepts like the written word, Edmund Curll was not going to let that get in his way.

Over the next few years, Curll went up against some of the greatest authors of the time – Matthew Prior, Jonathan Swift and Alexander Pope. He published what he called 'Keys' to famous works like Swift's *A Tale of a Tub* and *Gulliver's Travels*, which provided a loose explanation of these literary works to allow him to claim that he wasn't actually violating copyright, even though in the process he disclosed the entire storyline. He published wildly inaccurate biographies of famous people, happy to allow his authors to invent facts when the truth was not easily obtainable. He even invented the concept of the phantom poet – commissioning an author called Joseph Gay to write poems so that he could publish under the name J. Gay and pass them off as the works of the famous John Gay. So foul was his reputation that, in literary circles, the term 'Curllicism' became synonymous with literary indecency.

But Curll's most highly publicised battles were those he had with Alexander Pope. After more than a few minor skirmishes, he released one of Pope's anonymous poems against the express admonishments of the poet to the contrary. Enraged, Pope asked to meet him to discuss the incident. When they met, Pope vindictively slipped an emetic into Curll's drink, causing him to get so violently ill that he nearly died. Pope gleefully published a pamphlet carrying his account of the incident, informing the public that the notorious publisher was dead. Curll survived and swore he would get his revenge.

Curll then prepared a lurid version of the first Psalm that Pope had written some years ago and declared that he would be the future publisher of all of Pope's works. Shortly thereafter, when Curll published what he thought was a charitable biography of the former head of a local public school, Pope arranged for the students to wrap Curll in a blanket and beat him with sticks. Curll then published some of Pope's letters without authorisation, and Pope retaliated by having Curll figure prominently as a character of consistent ridicule and contempt in his 1728 masterpiece 'The Dunciad'.

The final straw came when, in 1737, Curll published five volumes of Pope's private letters, including the twenty-seven-year history of his correspondence with Swift – an exchange of letters that is, to this day, considered to be the finest example of that literary form, providing insights into the depth of the relationship between these two literary giants and its effect on the literature they had produced. In doing so, Curll had finally crossed a line. Pope took him to court.

When the court tried to apply the newly minted Statute of Anne, they were unable to find a way to get something as mundane as personal correspondence within the ambit of copyright. We have to remember that this was a time when copyright was still a new concept and it was not entirely clear what its boundaries were. The court struggled to punish Curll under copyright law, even though it was evident to them that an injustice had been done that needed to be addressed:

> I think it would be extremely mischievous to make a distinction between a book of letters, which comes out into the world, either by the permission of the writer, or the receiver of them, and any other learned work.[3]

Eventually, rather than extending the principle of copyright beyond what it was intended to cover, the court found a way to rule in favour of Pope by articulating a right to the privacy of correspondence. The court recognised that there was value in personal communications, and the mere fact that the letters were never intended to be published could not detract from that value:

> It is certain that no works have done more service to mankind than those which have appeared in this shape, upon familiar subjects, and which perhaps were never intended to be published; and it is this makes them so valuable.[4]

This was the first proper articulation by a court of the concept of privacy in personal correspondence, an idea that would be reaffirmed time and again,[5] giving authors the ability to restrain the publication of their letters by those who received them. While on the face of it this ruling seems to establish a proprietary right to personal writings, it merely confirmed the need to safeguard personal privacy and indicated that the courts were willing to step in to protect it.

This was the first of many cases that would pit the need for individual privacy against the desire to avail the benefits that this new technology could bring. The courts recognised that the invention of the printing press had democratised text, bringing knowledge and entertainment to every strata of society. But at the same time they understood that it had allowed booksellers like Curll to make public, in ways that had never been possible before, the most private of conversations.

While the value of privacy was slowly established over the course of many years, it would take a case that involved the English royal family itself to actually formalise the right.

Queen Victoria and her husband Albert occasionally, and for their own amusement, made drawings and etchings of subjects of private and domestic interest to them. These included portraits of their children, the Prince of Wales and the Princess Royal, other members of the royal family, personal friends and even their favourite dogs. Each of these etchings was transferred to copper and then printed – once again, only for their own private use – with the help of a private printing press that they had installed for that purpose. The plates of these etchings were kept under lock and key, though the impressions themselves were scattered across the Queen's private apartments around the country.

William Strange, an art dealer, somehow managed to obtain sixty-three of these impressions from a gentleman called Brown who worked in the royal family. Strange made it known that he intended to publicly exhibit these etchings even though he did not have the permission of the royal family. A law suit was brought and Strange was prevented from going ahead with the exhibition. Strange appealed to the High Court of the Chancery, claiming that since he had committed no fraud in obtaining these etchings – and, what's more, was unaware that there was any fraud involved – he should be allowed to go ahead with his exhibition.

Viewed through the lens of today, this seems to be an obvious breach of copyright, but at the time copyright was closely linked to publication and it was unclear whether there existed a right to protect an unpublished work under copyright. Works such as these personal etchings occupied a grey area in the law and so the court removed the case from the realm of property law, examining it instead under the law of breach of trust and confidence. Since the etchings were of a private character, it came to the conclusion that the impressions had been improperly obtained. The fact that Strange did not at the time believe the etchings to have been obtained through any impropriety was irrelevant so long as, on the contrary, he did not have any means to demonstrate that they had been properly obtained.

In his judgment, the Lord Chancellor said that given the effect that Strange's actions had on privacy, the court was going to rule in favour of the royal family, even in the absence of a clear statute that protected it. He went on to add a couple of sentences that made this judgment one of the foundations for the English law of privacy:

> In the present case, where privacy is the right invaded, the postponing of the injunction would be equivalent to denying it altogether. The interposition of this Court in these cases

> does not depend on any legal right; and, to be effectual, it must be immediate.[6]

With this, privacy was finally given the status of a right by the courts. In the decades to come, the right would solidify along the principles laid down in these two cases. As a result, the law of privacy in the United Kingdom evolved along the lines of breach of confidence and trust – principles that were subsequently expounded on and developed in a number of cases.

In the case of Ashburton v. Pape,[7] an injunction was sought to be brought against one Edward Pape, who had tried to use the letters between Lord Ashburn and his solicitor in bankruptcy proceedings. The court prevented him from doing so on the grounds that Pape had tricked the solicitor's clerk into giving up letters that he was under an obligation to keep confidential. In the case of Saltman Engineering Company v. Campbell,[8] the court found that 'the obligation to respect confidence is not limited to cases where the parties are in contractual relationship'. In doing so, the court established that breach of confidence was a remedy in tort and that parties did not have to have a formal contract. It allowed a breach of privacy to be exercised more broadly against a larger universe of people who shared a confidential relationship. In the case of Argyll v. Argyll,[9] the Duke of Argyll attempted to disclose evidence of his wife's polyamorous life to the press. The court prevented him from doing that, observing that 'there could hardly be anything more intimate or confidential than is involved in a marital relationship'.

Coco v. Clark[10] was a landmark case in commonwealth privacy jurisprudence, in which the court set out a three-part test for the breach of confidence: the information must have 'the necessary quality of confidence about it'; it 'must have been imparted in circumstances importing an obligation of confidence' and there must be an 'unauthorised use of that information to the detriment of the party communicating it'. This test ultimately became the touchstone on which English courts evaluated privacy in modern times.

The evolution of jurisprudence in England revolved around privacy violations that came about as a consequence of a breach of trust and reliance on trusted relationships. As a result, the law, as it developed in the UK, was designed to protect an individual against betrayal by confidants. It did not require the existence of a formal contract but did require demonstration of a relationship of confidentiality.

But this was a limited construction of the concept of privacy. What was required was a law that declared a right to privacy that could be invoked on its own against anyone who violated it, regardless of whether or not there was a relationship between the parties. For that to come to be, it would take the development of yet another technology on the other side of the Atlantic.

6
The Right

It was the end of the nineteenth century and the United States was suffering from the twin pressures of industrialisation and urbanisation. From 1800 to 1890, the population of the US had risen from 4 million to 63 million, with the greatest population growth being witnessed in the urban areas on the East Coast. Where in 1840 there were just twelve cities in all of the United States that had populations over 25,000, by 1890 that number had risen to 124, with some cities growing a hundredfold during those fifty years. People lived in 'overcrowded tenements ... and teeming slums',[1] thrust into such close proximity with each other that they could have no reasonable expectation of any sort of privacy.

These circumstances gave rise to an urban divide. The vast majority of the urban populace slept in relative squalor, but woke up alive to the promise that life in the vast oasis of opportunity offered – willing to put up with their current circumstances in the hope of getting the lucky break that would pull them out of mediocrity and into riches and fame. Thanks to the democracy of city streets, even the poorest slum dweller in New York City had the chance to see exactly how the rich and famous lived and all of them aspired, one day, to be able to live like that.

Much like Edward Curll in the UK over a century ago, the popular press of the day tuned into the aspirations of their readers and jumped in to fuel those dreams. They looked to fill their pages with stories of public figures and their lives in high society. In order to do this, they leveraged many of the new technologies that were coming into the mainstream at just about that time. Alexander Graham Bell had just invented the telephone, and the world's first commercial telephone exchange opened in Boston in 1877. By 1890, telegraph networks had been established around the country, creating a new world in which effective long-distance communication was the norm. These two technological advancements gave a huge fillip to the news industry, allowing papers to carry news of activities that happened further away and with greater proximity to the time of their occurrence than was earlier possible.

This resulted in a huge boom in the newspaper industry. Between 1850 and 1890, the number of newspapers in the United States grew from 100 to 900 and the readership grew from 800,000 to about 8 million.[2] Most of these newspapers were targeted at urban industrial workers whose thirst for gossip and the prurient details of the lives of the rich and famous had to be slaked daily.

It was at this time that George Eastman invented the Kodak portable camera and truly democratised photography:

> Photography was no longer the province of the professional and affluent amateur, but was practiced by thousands and thousands of people ... By 1889, the *New York Tribune* was able to report that amateur photography is rapidly approaching, if it has not already reached, the dignity of a 'craze'. The *New York Times* also reported a remarkable increase in the popularity of photography as a hobby.[3]

The fact that these cameras could be easily carried around, often concealed in the hands, made taking photographs far easier than it had ever been. Amateur photographers were becoming photojournalists, opportunistically taking pictures of the rich and famous when their guard was down and then selling those pictures to newspapers and magazines. Readers couldn't get enough of these newspapers, giving the press every incentive to push boundaries even more aggressively than ever before.

All of this created a new and particularly intrusive type of press corps that was armed with powerful new technologies and empowered by the protections guaranteed to them by the First Amendment in the American Constitution. It resulted in journalism that was geared to produce more and more salacious gossip every day in order to satiate the demands of the growing readership.

In a speech in 1886, President Grover Cleveland called the press the publishers of 'mean and cowardly lies that, every day, are found in the columns of certain newspapers, violate every instinct of America's manliness, and in ghoulish glee desecrate every sacred relation of private life'.[4] His chagrin was perhaps less out of a sense of presidential duty and more in response to the fact that the likeness of his wife Frances was being liberally used on product advertisements across the country without her consent. That said, it is always a mistake to invoke the ire of a sitting president, and it prompted, in very short order, one of the nation's first privacy laws to be enacted by the New York legislature, making it illegal to use an unauthorised likeness for commercial purposes.

It was into this charged environment that two young lawyers, recently graduated from Harvard and about to build a name for themselves as commercial lawyers in their own brand-new Boston law firm, decided to author, in the *Harvard Law Review*, an article on the right to privacy. One of them would go on to become one of the most highly regarded judges of the Supreme Court of the United States, responsible for handing down some of the most far-reaching judgments on the right to the freedom of speech and privacy ever delivered by the Supreme Court. The other lived a life of comparative obscurity, so consumed by the desire to protect the reputation of his family that it would eventually take a toll on his professional life, his friendship and his family. So much so that when he died, it was at his own hands – unable to live knowing that, after all he had done for his siblings, they suspected him of a conspiracy to deny them their rightful share in his family business.

Louis Brandeis was born into a family of Ashkenazi Jews, the youngest of four children who emigrated from their home in Prague to the United States after a series of political upheavals resulted in anti-Semitic riots in their hometown. Brandeis joined Harvard Law School when he was eighteen. His academic prowess and sheer grasp of complex nuances of law were so formidable that he graduated two years later at the top of his class, with the highest grade point average in the history of the school – a record that would stand for eighty years. The person he pipped to the finish line was his classmate and close friend through law school, Samuel D. Warren, with whom he would go on to found the law firm of *Warren and Brandeis.* It was so successful that it remains, albeit with a different name,[5] in continuous practice till today.

In the early years of their practice, Warren and Brandeis jointly published three articles in the *Harvard Law Review*. The third article, published in 1890, was simply entitled 'The Right to Privacy' but has been called the 'most influential law review article of all'[6] and an 'outstanding example of the influence of legal periodicals upon the American law'.[7] Roscoe Pound, an eminent jurist in his own right, said that it did 'nothing less than add a chapter to our law'.[8] Its repercussions were felt around the world and continue to influence judgments even to this day. When the Indian Supreme Court was deciding in 2017 whether or not the country had a fundamental right to privacy, the judges were unanimous in recognising the role that this article played in conceptualising privacy as a modern jurisprudential concept.

Far more interesting than the article itself, however, is the reason why it was written.

Samuel Warren, Jr was the oldest of five children. While each of his siblings – Henry, Cornelia, Edward and Frederick – turned away from a life in 'high society', Sam aggressively pursued it. His first step was marrying well – into the upper crust Bayard family, whose 'old money' and political power helped him rise to academic and social success in Harvard. This unlocked, for him, membership into some of the most important Boston clubs and cultural institutions, and helped him build a successful law firm. However, despite marrying above his station, he remained on the social periphery as it was his vivacious wife Mabel and the rest of the Bayard clan on whom the social spotlight was focussed.

The Warren family, by comparison, couldn't have been more different, steeped as they were in secrets that were scandalous by the standards of the time. As much as Samuel Warren's prospects had been improved by marriage, his family, which continued to remain 'barely compatible individualists – not easily compatible with other, milder people, let alone with each other',[9] was a source of stress and a millstone around the neck of a young man trying to climb up the social ladder.

Of particular concern was his brother – Edward Perry Warren – who, upon achieving adulthood, became aware that he was attracted to persons of the same sex. At that time in American history, scientists had begun to cruelly categorise men like him as 'sexual deviants', 'pederasts' and 'inverts'. Most middle-class Americans of the time associated same-sex intimacy with the notion of 'degeneracy' – something far more morally dangerous than the temporary moral deterioration previously associated with episodic sodomy. Many believed homosexuality was just one of the many signs of the social 'disorder' in American society – and thought of it as a modern curse on a par with the growing women's rights movement, the rapidly rising divorce rates, and the increase in prostitution and venereal disease in urban areas. As a result, the 1880s saw a rapid rise in the conviction of sexually deviant men, including those who came from respectable families or were socially prominent figures like Oscar Wilde.

Ned Warren never made much of an effort to present himself as masculine in the way that was expected of men of the time. Instead, he enthusiastically and overtly embraced his sexuality, vocally identifying himself as a Platonic 'aesthete'. After his father's death in 1888, he spent the enormous annual stipend he received as inheritance on a Sussex mansion dedicated to the appreciation of art and sensuality in the ancient Greek tradition, in which he and a group of like-minded gay men lived communally.

Samuel Warren, as the new head of the family, was particularly protective of Ned – the most vulnerable of the Warren 'children' – given how his mannerisms, personal interactions and aesthetic interests clearly marked him as a presumptive homosexual. Perhaps this anxiety was enhanced by the fact that Sam was mindful of the prominence of the family he had married into, and was concerned that the increasingly inquisitive and scandal-seeking press of the time would not hesitate to lay bare the dark secrets of his brother's sexuality should they get whiff of the scandal.

It is very likely that it was a fear of these deeply personal consequences that fuelled Warren's desire to articulate a strong legal basis with which he could defend his brother against the depredations of an aggressive press.[10] Little did he know that the paper he wrote would be so influential that it would become the cornerstone of privacy law the world over and would remain relevant over a century and a quarter after its writing.

To be clear, Warren and Brandeis were by no means the first people to think about the concept of a legal right to privacy. James Madison, one of the architects of the American Constitution, had tried to articulate something like a right to privacy during his day but found it difficult to express himself in the language of the times. In those formative days of modern democracy, there was no construct within which unsubstantial notions such as privacy could be presented. Madison had to frame his thoughts in the context of the right to property, and he struggled to squeeze into that construct the notion of human conscience. He spoke of it as a sacred form of property that needed to be given the status of a natural and inalienable right:

> A man has property in his opinions and the free communication of them...He has an equal property interest in the free use of his faculties and free choice of the objects on which to employ them. In a word, as a man is said to have a right to his property, he may be equally said to have a property in his rights. Where an excess of power prevails, property of no sort is duly respected. No man is safe in his opinions, his person, his faculties or his possessions...
>
> Conscience is the most sacred of all property; other property depending in part on positive law, the exercise of that being a natural and inalienable right. To guard a man's house as his castle, to pay public and enforce private debts with the most exact faith, can give no title to invade a man's conscience which is more sacred than his castle, or to withhold from it that debt of protection, for which the public faith is pledged, by the very nature and original conditions of the social pact.[11]

Many jurists have since written that his concept articulated a right to personal space without being able to properly express it in terms that would be relevant in the context of modern technology. Justice Thomas Cooley, in his *Treatise on the Law of Torts*, was perhaps the first person to spell it out in language that is used today, saying that 'the right of one's person may be said to be a right of complete immunity; the right to be alone'.[12]

As a result, there was already some recognition of the concept of personal privacy as a distinct right towards the end of the nineteenth century, and the phrase 'right to privacy' was not entirely unheard of. However, with the exception of a few laws that made eavesdropping a crime, privacy was little more than a by-product of equitable remedies designed to address other wrongs. If, in 1890, you wanted an injunction from an American court, there was almost no way in which you could obtain it unless you could prove an injury to property. Which meant that if someone's personal affairs – including those of a deviant homosexual brother – were about to be made public by a newspaper, there may not have been any real cause of action for an injunction.

The other reason why it was so hard to clearly articulate a right to privacy against the press was because of the almost unbridled freedom that was guaranteed to them under the First Amendment of the US Constitution. In the words of Justice Cooley:

> The constitutional liberty of speech and of the press, as we understand it, implies a right to freely utter and publish whatever the citizen may please and to be protected against any responsibility for so doing, except so far as such publications, from their blasphemy, obscenity, or scandalous character, may be a public offense, or as by their falsehood and malice they may injuriously affect the standing, reputation, or peculiarly interests of individuals.[13]

It was this powerful formulation that gave the press the ability to publish with impunity, hunting down gossip and personal information of the rich and famous and serving it up for the enjoyment of the masses. It was the principal argument that the newspapers used to win cases that were levelled against them that sought to curb their publications.

Brandeis and Warren recognised the impact that the new technologies of the day were having on the right to be left alone. They believed that the manner in which these new technologies were impinging on one's personal privacy deserved special attention even if, in the process, this called for the application of some fetters on the freedom of the press. They observed that:

> Recent inventions and business methods call attention to the next step which must be taken for the protection of the person, and for securing to the individual what Judge Cooley calls the right 'to be let alone'. Instantaneous photographs and newspaper enterprise have invaded the sacred precincts of private and domestic life; and numerous mechanical devices threaten to make good the prediction that 'what is whispered in the closet shall be proclaimed from the house-tops'. For years there has been a feeling that the law must afford some remedy for the unauthorized circulation of portraits of private persons... [14]

They were particularly concerned about the extent to which these new technologies were beginning to exacerbate the already considerable intrusiveness of the press. They noticed that gossip, which was previously 'the resource of the idle and vicious', had now become a trade that was being 'pursued with industry as well as effrontery' to satisfy a prurient taste. While the complexity of modern life had rendered necessary a 'retreat from the world', and man, having become more sensitive to publicity, was feeling an increased desire for solitude, modern inventions had made it easier for that privacy to be invaded, subjecting him to the sort of mental pain and distress that was far greater, in their view, than could be inflicted by bodily injury.

Of particular concern were the repercussions that portable cameras were having on privacy. They knew that the law relating to confidentiality offered them some recourse but were not satisfied with basing their defence on that line of argument. As we have seen in the context of the development of law in the United Kingdom, a jurisprudence that is based on the notions of confidentiality requires the existence of some sort of a relationship between the parties. No such relationship could be established when privacy was being invaded by strangers equipped with cameras who had no connection whatsoever with the subject of the photograph.

How was the law to redress injuries caused by this new invention? While breach of privacy does bear some resemblance to the wrongs dealt with under the laws relating to slander and libel, defamation only offered redress for injury caused to the individual in the context of his external relations with others in society – injury that occurs because he has been lowered in the estimation of his fellows. It didn't fully redress the injury caused to his feelings.

Some of the principles of law that had been developed in the UK in the field of intellectual property could be drawn upon. General common law could also possibly form the basis for such a remedy. These laws allow individuals to determine the extent to which their thoughts must be communicated to others and follow the principle that, except on the witness stand, no man should be compelled to express his thoughts. They stipulate that when he does communicate them, it is up to him to establish the limits within which they are to be publicised, and that no one has the right to publish the private thoughts and writings of another without his consent except when he himself makes his thoughts public.

These rights were not based on the right to profit from publication, but instead drawn from the 'the peace of mind or the relief afforded by the ability to prevent any publication at all'.[15] Warren and Brandeis seized on these concepts to refine their argument for privacy, using illustrations similar to those we have come across while studying the development of privacy in the United Kingdom – even though, in their hands, these principles were spun to achieve an entirely different outcome. They gave the example of a man's entry in his diary about dinner with his wife on a certain day, and stated that no one could legitimately publish this sort of a personal diary to the world even if they could prove that they had legally obtained access to it. What prevents publication, argue Brandeis and Warren, is 'not the intellectual act of recording the fact that the husband did not dine with his wife, but that fact itself'.[16]

They used this example to articulate a subtle difference between the intellectual value of an intangible product – which is protected as a literary composition under copyright law – and the domestic occurrence of a personal activity, which needs to be separately protected. They pointed out that the protection that is extended to thoughts, sentiments and emotions is an example of the enforcement of the more general right of the individual to be left alone.

> It is like the right not to be assaulted or beaten, the right not to be imprisoned, the right not to be maliciously prosecuted, the right not to be defamed. In each of these

> rights, as indeed in all other rights recognized by the law, there inheres the quality of being owned or possessed – and (as that is the distinguishing attribute of property) there may some propriety in speaking of those rights as property. But, obviously, they bear little resemblance to what is ordinarily comprehended under that term. The principle which protects personal writings and all other personal productions, not against theft and physical appropriation, but against publication in any form, is in reality not the principle of private property, but that of an inviolate personality.[17]

Using this argument, they articulated the principle that could be invoked to protect the privacy of individuals from invasion by the press. They recognised that a number of cases had already been decided based on the breach of trust implicit in contractual principles. However, given the growth of modern technologies that offered far greater opportunities for the perpetration of similar wrongs, they argued that there was a need for these protections to be based on a much broader foundation. During the early days of photography, no one's picture could be taken unless they 'sat' for the photograph. Under those circumstances, it was possible to secure adequate protection against the improper circulation of his or her image by relying solely on the law of breach of trust, because a photograph could only be taken in a studio with the express consent of the subject and under the conditions that that person stipulated. Now that portable photography made it possible to take pictures surreptitiously, protections based on contract and trust were clearly inadequate.

For these reasons, they argued, the individual needed some way to exercise his right to be left alone against the world – as a right *in rem*. Accordingly, the principle that they felt should be applied to protect personal writings and any other similar productions was the right to privacy – the right that every private individual has to protect himself from portraiture, or to protect discussions about his private affairs from being shared with the public.

In the years immediately following the publication of the article, a number of courts in the United States referred to this concept in their judgments – and, even if they did not expressly articulate a right to privacy as laid out in the article, they adopted many of the concepts broadly discussed by Brandeis and Warren in the paper.

The first decision that referenced the article – Schuyler v. Curtis[18] – involved neither the press nor a publisher. This case revolved around the erection of a statue of the late Mrs George Schuyler at the Colombian Exposition in Chicago without the consent of anyone in the Schuyler family. Judge O'Brien, while granting an injunction, tried to assess, in accordance with the line of thinking laid down in the article, whether the late Mrs Schuyler was in fact a public character and finally concluded that, because she was not, the family was entitled to privacy. In the process, he specifically referred to the Brandeis-Warren article as his basis for the extension of tort law to the realm of privacy and quoted liberally from the article, stating that it was a piece of writing that 'well deserves and will repay the perusal of every lawyer'.

When a law suit was brought against the Rochester Folding Box Company, for printing thousands of copies of the portrait of a young Ms Roberson and using them, without her consent, as advertisements for flour, the court held that principles of equity could be extended to protect more than just property rights and that the plaintiff had the right to an injunction based on the right of property that the plaintiff had in her own body.[19] However, the New York Court of Appeals, by a narrow 4-3 majority, overturned this decision on the grounds that no precedent existed for this sort of an extension of the law of tort. The court felt that establishing a new tort action like this would inundate the court with a number of petty and absurd claims of invasion of personal privacy such as a comment on 'one's looks, conduct, domestic relations or habits'.[20]

This decision was widely criticised and became the focal point of a much broader debate on privacy, resulting in vociferous calls for a more clearly enunciated right to privacy. The people of New York were no longer willing to meekly tolerate the interference by the press into their personal affairs. This was, in many ways, the United States' 'Aadhaar' moment.

The resulting outcry led directly to the enactment of a law that made the use of a person's name, portrait or likeness without their consent a misdemeanour. Eventually, several other states enacted similar laws, firmly bringing the concept of privacy into the statute books.

The first decision that actually called out invasion of privacy as a distinct cause of action was the 1905 case of Pasevich v. New England Life Insurance Company,[21] which dealt with the non-consensual use of the plaintiff's picture in a newspaper advertisement along with a promotional statement incorrectly attributed to him. The Supreme Court of the State of Georgia held that there was such a thing as a right to privacy in natural law, and that an infringement of privacy was a direct invasion of that legal right. The court pointed out that publishing someone's photograph without his consent cannot be a form of expression of a sentiment or an idea that should be entitled to receive constitutional protection.

> Those to whom the right to speak and write and print is guaranteed must not abuse this right, nor must one in whom the right to privacy exists abuse this right. The law will no more permit the abuse by the one than by the other. Liberty of speech and of the press is and has been a useful instrument to keep the individual within the limits of lawful, decent and proper conduct; and the right of privacy may well be used within its proper limits to keep those who speak and write and print within legitimate bounds of the constitutional guarantees of such rights. One may be used as a check upon the other, but neither can be lawfully used for the other's destruction.

This was the decision that finally elevated the concept of the right to privacy as articulated in Brandeis and Warren's paper to sit on par with the protections offered to the press under the right to freedom of speech. It gave other courts in the country a judgment from a well-respected state Supreme Court that they could rely on. Within six years, the principles of the right to privacy, as set out in the article, were being cited by courts around the country. New Jersey issued an injunction against the use of Thomas Edison's name and picture. Kentucky and Missouri shortly followed suit, with similar injunctions in other cases that involved the invasion of privacy.

Eventually, in 1928, the issue came up before the Supreme Court of the United States and featured prominently in the case of Olmstead v. United States.[22] This was a case in relation to whether evidence obtained through the wire-tapping of private telephone conversations was a violation of the Fourth and Fifth Amendments of the US Constitution. Louis Brandeis, now himself an associate justice of the Supreme Court, had the opportunity to write the dissenting opinion. He noted that 'subtler and more far-reaching means of invading privacy have become available to the Government. Discovery and invention have made it possible for the Government, by means far more effective than stretching upon the rack, to obtain disclosure in the court of what is whispered in the closet.' He argued that:

> The makers of our Constitution undertook to secure conditions favorable to the pursuit of happiness. They sought to protect Americans in their beliefs, their thoughts, their emotions and their sensations. They conferred, as against the Government, the right to be let alone – the most comprehensive of rights, and the right most valued by civilized men. To protect that right, every unjustifiable intrusion by the Government upon the privacy of the individual, whatever the means employed, must be deemed a violation of the Fourth Amendment.

Brandeis's dissenting opinion in the Olmstead case has since been invoked by the Supreme Court in a variety of criminal procedure decisions over the next few decades.[23] By the 1960s, thanks to Joseph McCarthy and the extreme threat that his radical philosophies posed to civil liberties, the principles behind the right to privacy began to move past the narrow limitations of criminal procedure and, when the Supreme Court ruled on the matter of Griswold v.

Connecticut,[24] were applied in relation to the constitutional validity of a Connecticut law prohibiting the sale and distribution of contraceptives to married couples. Justice William Douglas, who delivered the majority decision, held that such a law violated the right to marital privacy that emanated from the 'penumbras of the fundamental constitutional guarantees of the rights in the Bill of Rights', which together create 'zones of privacy'. This was the most expansive articulation of the right to privacy yet and would be used in a number of judgments that established the law of privacy.

Samuel Warren wanted nothing more than to protect his younger brother from being exposed as a homosexual at a time when being gay was punishable with the most cruel torments. The press was answerable, at the time, to no one and constantly on the lookout for a scandal. They had access to powerful new technologies with which anyone could take a photograph of anyone else without their consent and without a care as to how, by doing so, it would harm their privacy. It was this flagrant abuse of a person's right to be left alone in the face of new technologies that was uppermost on his mind as he finalised his seminal paper.

The paper was timely and tapped viscerally into public sentiment. It came at a time when the upper echelons of society were themselves growing increasingly frustrated by the manner in which technology was insinuating itself into the private sphere and courts were struggling to find ways in which to prevent these violations with the tools they had in hand. As a result, the concepts set out in the article gained wide acceptance. They offered courts a new way to think about privacy, taking it outside the realm of confidentiality as it had existed till this time and allowed them to create a right where none previously existed.

With that, the notion of a right to privacy had found its way into the constitutional jurisprudence of the United States. It went on to spread across the world, being quoted in judgments as far away as India, remaining as relevant over a century after it was first articulated. And while different countries have charted different paths on their road to privacy, almost all of them owe their origins to this paper.

There is an epilogue to the story of Ned and Samuel Warren. While there is no evidence that any harm befell young Ned on account of his lifestyle, the work that Samuel Warren started on his brother's behalf eventually came to benefit the community that his brother represented. Through a series of cases, each building upon the precedents set by those that went before it, the jurisprudence of privacy developed to protect the LGBTQ community.

In Eisenstadt v. Baird,[25] the principle of the fundamental right to privacy was extended beyond the marital home to include unmarried persons when the court invalidated a law prohibiting the distribution of contraceptives to unmarried persons. This decision was followed in the seminal case on self-determination – Planned Parenthood of Southeastern Pennsylvania v. Casey[26] – and subsequently in the landmark gay rights judgments in Lawrence v. Texas[27] and United States v. Windsor.[28]

This series of judgments eventually led to Obergefell v. Hodges,[29] a 2015 decision of the US Supreme Court in which the court, relying on the principles in the paper by

Brandeis and Warren, held that everyone has the fundamental constitutional right to marry another person of the same sex.

It took 125 years but, finally, the paper that Samuel Warren wrote to protect his gay brother was instrumental in securing a crucial fundamental right for the entire gay community – one that would do more to bring them within the formal fold of normal society than anything had done before.

PRIVACY 2.0

7

The Currency of Information

In the early days of the American republic, the only means to send information across the vast distances of the continent was by post. But even though the postal network offered an efficient way to carry messages to the far corners of the United States, a sealed envelope did little to prevent someone who was determined enough to do so from gaining access to the contents of the letter he was delivering. The problem was so bad that Benjamin Franklin, who was then in charge of the colonial mail, forced his employees to take an oath that they would not open the mail.

The American Congress passed a law in 1782 that prohibited mail from being opened en route – one of the early regulations on privacy in the US – but the problem persisted and personages as eminent as Thomas Jefferson, Alexander Hamilton and George Washington complained bitterly about the lack of privacy in correspondence.

Jefferson blamed the 'infidelities of the post office and the circumstances of the times' for his disinclination to write fully and freely. Ralph Waldo Emerson wryly noted that it was unlikely that 'a bit of paper, containing our most secret thoughts, and protected only by a seal, should travel safely from one end of the world to the other, without anyone whose hands it had passed through having meddled with it'.[1]

Eventually, Congress enacted a more robust statute in 1825 which stipulated that:

> Whoever takes any letter, postal card, or package out of any post office or any authorized depository for mail matter, or from any letter or mail carrier...before it has been delivered to the person to whom it was directed, with design to obstruct the correspondence, or to pry into the business or secrets of another, or opens, embezzles, or destroys the same, shall be fined...or imprisoned.[2]

With the invention of the telegraph, messages could be carried further in a much shorter time. Since there were far fewer points at which these messages could be intercepted, this should have been a far more secure method of communication across long distances. But everyone soon realised that the technology was far from infallible. During the American Civil War, both the Union and Confederate armies developed wire-tapping technologies so that they could overhear each other's long-distance chatter to try and figure out battle plans and troop movements. This drove home to the governments of the day the strategic importance of being able to control the flow of messages and, as soon as the war was over, the US Congress sought access to the telegraph operations of Western Union.

This attempt by the government to create a back door into the channel through which private communications travelled was met, in much the same manner as it is today, with strong outrage. A *New York Times* editorial decried the government's attempt to control telegraph operations as 'an outrage upon the liberties of the citizen'. The *New York Tribune* wrote that the seizure of telegrams violates 'every man's sense of his right to the secrets of his own correspondence'. The *New York Sun* said that 'the idea that every curious and prying legislative committee may cause to be spread before the public everything that has been sent over the wires will be hateful and repulsive to the people in general'.[3]

Eventually the courts stepped in, quashing several subpoenas for the contents of telegram messages, comparing them to letters that were already protected under law. The Missouri Supreme Court, while setting aside one such request, held that such 'an inquisition, if tolerated, would destroy the usefulness of this most important and valuable mode of communication'.[4] Eventually, bowing to the pressure from a concerted opposition, the legislatures of various states of the Union passed laws that prohibited the disclosure of telegraph messages – along much the same lines as they had previously prevented the reading of postal messages.

The fact that the government has the power to take control of our personal communications has built up in us an innate distrust of the state and the government machinery. The ability to intercept communications remains, to this day, one of the prerogatives of the state and an example of some of the most egregious violations of personal privacy of the modern day. Almost every country in the world has written these powers into its statute books, and since they all insert exceptions to privacy on the grounds of national security, the limited role that law or the judicial system has to play in the context of the use of interception is to evaluate whether the executive has exceeded its authority or not. This mistrust of the government on matters of surveillance and interception of communication continues to the present day – and has been exacerbated in recent times by the revelations on WikiLeaks and from Edward Snowden.

But as personally invasive as active surveillance can be, the harm that could result from passive surveillance is far more insidious. Today, data is collected from us in ways we do not fully comprehend. Information is being analysed to create profiles of us that help corporations and governments get a better idea of who we are as persons, ostensibly so that they can improve their services and provide newer and better products. Through this process, vast databases of personal information have been built and are continuously added to. They can be drawn upon as required, and contain, in the form of records of our movements, our choices and our behaviour online, an image of us that accurately outlines our life and personality. They represent a form of surveillance that is far more persistent and, consequently, has far deeper consequences than active surveillance. Of the many ways in which our privacy can be violated today, this has the potential to be most deeply invasive.

The role of personal data in businesses owes its origins to developments in the eighteenth century among the tradespeople who operated in the large commercial capitals of the western world. In those days, you got credit based on familiarity. When towns were small, shopkeepers and tradesmen knew all their customers personally. From local gossip and common chatter, they had a fair idea how their customers' businesses were doing, and the reputation of their families. Almost everyone knew where everybody else lived. As a result, extending credit was a social function of a business done on trust.

As towns grew larger, urban centres of trade and commerce began to witness an influx of strangers. Traders were forced to deal with more and more people they didn't know and whose creditworthiness they couldn't assess. Some of those who, in good faith, extended credit to these strangers lost money as a good number of them were nothing more than cunning thieves taking advantage of a system based on trust.

Tradesmen began to band together for their own protection. In London, the Society of Guardians for the Protection of Trade against Swindlers and Sharpers was established in 1776 in order to communicate amongst the members the name and description of any person who was unfit to trust. By 1812, it had over 550 members spanning nearly every trade in the city of London.[5] Similar organisations cropped up in Liverpool, Bath, Leicester, Yorkshire, Glasgow and Aberdeen. Most of these organisations were required in their bye-laws to maintain communications with other such organisations in other cities so as to share information about those who made it a business of swindling tradesmen.

Across the Atlantic, the issues were no different. The American towns that were once small enough that a stranger would stand out were now bustling metropolises. Businesses were struggling to follow the same trust-based approach that they had been using for so long. Into this context stepped

Lewis Tappan, a Massachusetts businessman better known for his role in freeing the African slaves on the *Amistad*. He created a new business model that would so radically transform the way in which businesses operated that its repercussions are felt to this day.

As a strict Calvinist, Lewis Tappan began his career by insisting on transacting in cash. He believed in strictly following the Biblical strictures against lending money and charging interest. One will never be able to tell for certain whether it was because of this puritanical approach to commerce or the sheer poor timing of his (rather speculative) investments in woollen and cotton mills, but Tappan soon went bankrupt.

When he did recover, he seemed to have learned his lesson and, with the help of his brother Arthur, decided to adopt a more flexible (and secular) approach to business. He began to cautiously extend credit to his customers. However, unable to forget his Christian upbringing, he could not help but keep detailed records of their re-payment history and general creditworthiness. In time, as these records grew, his fellow merchants began to ask him for advice, checking on whether the customers they were dealing with were good for their word.

Ever the businessman, Tappan spotted an opportunity. He began to publish his credit ratings in digest form, selling it to tradesmen all over town, allowing them to use it as a ready reckoner of an individual's credit. He was scrupulously careful to be accurate, providing ratings solely based on an individual's ability to pay and his assessment of how long they would take. He devised a simple alphabetic rating system, with a key that explained exactly what an A, B or C meant.

His fellow tradesmen loved it. If a new customer crossed their threshold, they could simply look his name up in Tappan's book and, based on the clearly described rating, decide for themselves whether it was worth doing business with him or not. In time, demand for this sort of service was felt in other cities in New England. Tappan leveraged his abolitionist connections to establish a network of correspondents whom he paid to generate up-to-date, comprehensive records of people in their communities. These correspondents included a young Illinois lawyer called Abraham Lincoln and a Midwestern storekeeper, Ulysses S. Grant. By 1845, his firm, the Mercantile Agency, had offices in Baltimore, Boston and Philadelphia, and businesses all over the East Coast were using his credit reports to do business with people they had never met.

As the business of credit reporting caught on, many others jumped on to the bandwagon. It was a relatively easy business to operate – all you needed was an easy-to-understand rating system and the ability to assess whether or not a person had social character or was a business risk. However, in order to stand out among the competition, credit rating businesses began to collect more and more information about the individuals they assessed – information that, while not directly financially relevant, was nevertheless indicative of their ability to pay.

For instance, married men were considered to be more responsible than bachelors and therefore rated to be more likely to repay debts. Anyone who had an injury or suffered from a medical condition was rated as having a diminished ability to work, representing a greater credit risk than someone who was able-bodied. As demand for their services increased, credit rating agencies grew more and more ingenious about these proxies, eventually including hearsay and social opinions into their analysis.

As repugnant as this might seem when you think about it, this is not very different from the way things have always been. We seek out the opinions of those we trust before entering into new commercial relationships. It is our way of reassuring ourselves about those we do business with. As populations grew, it was becoming increasingly difficult to effectively gather this sort of social intelligence. All that the credit rating businesses were doing was offering a new way in which to achieve the same results.

That said, any business that analysed personal data in order to build a profile of people was, by the very nature of the business, violating the privacy of the people whose data they were processing. The right to privacy set out in the Warren-Brandeis paper and the many judicial decisions that subsequently affirmed it would have applied to them, and unless they managed to find some form of legal safeguard, they ran the risk of prosecution. In order to protect themselves, they eventually turned to the original legal construct on which privacy was based – the protection of confidential information that had been the basis for the development of the concept of privacy in the UK.

Now, every time any sort of personal data was collected from an individual, consent was obtained through standard form language describing a set of purposes broad enough to cover all possible uses to which that data could be put. Most customers didn't think twice about the consequences of providing that consent, incapable of appreciating how parting with this data could affect their personal privacy. But each item of data they provided eventually found its way into the hands of credit rating agencies, that added every little nugget of information they could gather to the information they already had, improving and refining the profiles they were building.

When computers and networked databases began to insinuate themselves into commercial life, the volume of information that these organisations could collect and process increased exponentially. With greater data storage capacity, more and more details about a person's life could be stored – until these agencies had almost as much information about the people they were tracking as they did themselves. With the improvement in processing power and advanced computation technologies at their disposal, these companies began to identify patterns in the data that had until then remained unseen, offering further novel insights into the people they were profiling.

Credit rating agencies today control some of the most detailed databases of personal information on the planet. They represent some of the most accurate profiles of creditworthiness and, as a result, can dramatically affect an individual's ability to start a new business, buy a new car or take out a mortgage on a home. Since the information that these agencies possess goes much further than financial data, these reports could also affect a person's ability to get a job and have other such non-financial consequences.

As reliance on the information contained in these reports has grown, we have begun to believe that they represent such an accurate assessment of a person's financial worth that we now trust them over any evidence that the applicant presents to the contrary. As a result, loss of identity has become a huge problem in most of the Western world. When Michelle Brown, a bank employee, deposed before the US Judiciary Subcommittee on Technology, Terrorism and Government

Information, her testimony was a chilling reminder of the devastating consequence of identity theft:

> Over a year and a half from January 1998 through July 1999, one individual impersonated me to procure over $50,000 in goods and services. Not only did she damage my credit, but she escalated her crimes to a level that I never truly expected: she engaged in drug trafficking. The crime resulted in my erroneous arrest record, a warrant out for my arrest, and eventually a prison record when she was booked under my name as an inmate in the Chicago Federal Prison.[6]

As data systems improve in accuracy to the point where they are rarely wrong, we will begin to trust them over the evidence of our senses. We have already come to believe that the data never lies – so much so that many of our decisions are based wholly on the data that is presented to us. Sometimes that data does lie, either because of input errors or because the analysis of the facts is faulty, but if we have come to trust in the infallibility of our data systems, we will stop applying our credibility filters to the data presented to us and all these mistakes will slip through.

When Lewis Tappan started the Mercantile Agency, there is no way he could have imagined what his fledgling business would grow to become. If he had, perhaps he might have done things differently. One of the early choices he had to make was how exactly to monetise the information he had under his control. He could have organised things so that he sold the individuals he was rating a certificate endorsing his opinion of their credit rating – a certificate that they could subsequently present whenever they needed to demonstrate their creditworthiness. This option would have ensured that personal information remained within the control of the individual to whom it pertained, allowing them to decide whom to share it with and for what.

Instead, he chose to create a directory in which he recorded the credit score of every person he had assessed, selling it to businesses which could use this book as a reference before extending credit to strangers. His choice of this second option probably had to do with the fact that it was likely to be more lucrative to sell his digest to businesses than certificates to customers, since the former had a real incentive to ensure that they knew the credit rating of the people they were selling to. Be that as it may, it is thanks to this initial business choice that personal information has, since then, been treated as a commodity that is collected and processed by businesses and used in ways that are beyond the ability of the individual to influence.

Lewis Tappan sold his Mercantile Agency to his chief clerk, Benjamin Douglass, in 1849 so that he could focus full-time on ridding America of slavery. He fought tirelessly for this cause and was able, within his lifetime, to see the Emancipation Proclamation issued by his one-time correspondent and then President of the United States, Abraham Lincoln. Douglass transferred the company to his brother-in-law, Robert Graham Dun, who renamed the business R.G. Dun & Company. In 1933, the company merged with its main rival, Bradstreet, to form the firm Dun & Bradstreet, a company that exists to this day and is possibly the largest credit reporting company in the world. Its survival is a testimony to the increasing importance of data in our modern context.

Tappan's design choice continues to affect the way data is collected and processed. Since he first started issuing credit ratings, the scope of the business expanded beyond merely providing credit information to helping businesses develop various personalisation tools that allow them to tailor their services to more specifically suit our personal preferences. As a result of that design choice, a tremendous asymmetry has developed between data collectors and data subjects.

This has resulted in the development of data protection laws that seek to limit the manner in which data can be used. These laws range from a simple set of principles intended to broadly regulate the manner in which data is collected and used to the complex, prescriptive legislation we have today. These are the frameworks within which data collectors around the world have to operate while collecting and processing personal information from their users (the data subjects). For the most part, they comprise increasingly complex sets of rules that corporations are obliged to follow but which at the same time impose upon the data subjects an obligation to be mindful of what they allow data controllers to do with their data.

The earliest data protection law grew out of the report of the HEW Committee in the United States, which recommended the establishment of a Code of Fair Information Practices based on Fair Information Practices Principles (FIPPS). This Code described how personal data should be handled, stored and managed with the objective of ensuring fairness, privacy and security in the context of new technologies.

In 1980, the Organisation for Economic Cooperation and Development (OECD) issued a set of guidelines that were heavily inspired by FIPPS, with the intention of providing a framework that OECD members could use to prepare domestic privacy laws that upheld the fundamental principles of human rights while at the same time allowed cross-border data flows. The OECD Guidelines were hugely influential and formed the basis for the European Directive 95/46/EC, the 2004 Asia-Pacific Economic Cooperation Framework as well as the privacy laws of Australia, New Zealand and Japan.

Data protection law today is a complex area of legal specialisation. Global corporations have large teams focussed solely on compliance with the many privacy laws that apply to them. Countries around the world are constantly tinkering with their legislation, constantly trying to reorganise their regulations to deal with the implications of modern technologies. But as much as these jurisdictions have had decades of experience regulating personal data, the approach that they follow when it comes to matters of data protection is heavily influenced by Lewis Tappan's original design. This approach continues to affect the manner in which we regulate new technologies such as big data and artificial intelligence.

8
Meanwhile, in India...

One of the reasons why India has been somewhat isolated from the evolution of privacy jurisprudence is that it has never really felt the initial impact of these new technologies in quite the same way as the rest of the world did. At the time when the portable camera and the telegraph first began to be widely used in the US and elsewhere in the world, India was still a far-flung colonial outpost of the British Empire. While these new technologies did eventually make their way to Indian shores, it wasn't until much after they had been deployed in the West and their impact on privacy was already well understood. As a result, when they were deployed in India, the laws that governed them already had safeguards built in. Most of the colonial-era legislation that is still in use in India (like the Indian Penal Code and the Indian Telegraph Act) contains statutory provisions designed to address privacy concerns or to provide exceptions to allow legal interception

– evidence that our current jurisprudence was based on legal frameworks that were transplanted with deliberation from another land.

That said, even when they finally did come to India, these technologies were used almost exclusively by the British East India Company for their own purposes. When they were eventually allowed to be used by the 'natives', I doubt that the threat to personal privacy was ever a matter of much concern, given that at that time Indians' liberties were already being curtailed by the colonial overlords. As a result, we in India never got to relate violations of personal privacy to incursions by technology in the same way as was the case in other countries in the world.

It is this fundamental difference in India's exposure to the personal consequences of the deprivation of privacy that has given rise until recently to our laissez-faire attitude to the technologies that clearly have an effect on our personal liberty. It is the reason why the Indian government was able to, virtually without objection, conceptualise, create and implement a project as ambitious in scope as Aadhaar, where elsewhere in the world similar projects have come to a standstill. In many ways, Aadhaar was the first technological shift that has had a direct impact on personal privacy that we have felt for ourselves in India. As much as we might be outraged by it, we have to appreciate that this reaction is no different from opposition to technologies that have come before it – the portable camera and the telegraph.

At the same time, it might be this very difference in our national experience of privacy that gives us a unique opportunity to think afresh about how to regulate privacy in the context of data. Around the world, regulators are struggling to understand how to unlock the value of data technologies while still protecting privacy. In this, they are constrained to follow the path that they have been on since they first began to regulate personal privacy – at a time well before the benefits of a data-intensive world were evident. We have no such path dependence, and as we set out to formulate from scratch our own privacy jurisprudence in the context of uniquely Indian technologies like Aadhaar, we have the freedom to chart a new path – one that is more aligned with the data-driven world that we find ourselves in.

9

Early Thoughts on Privacy

There is an impression, perhaps most deeply held by us ourselves, that Indians care not a whit about personal privacy. The example most frequently trotted out to support this proposition is that quintessential Indian cliché – the joint family, as if the fact that multiple generations living cheek by jowl with each other is all the evidence we need to support the fact that we can make do with far less privacy than the nuclear families of the West. Much is made of how we casually share intimate details of our private lives with relative strangers, happily dishing out personal information to whosoever asks for it without worrying about the consequences on personal privacy.

While there is some truth in all of this, it is far from accurate to say that Indians lack an appreciation for the basic concepts of personal privacy. Indian civilisation has probably understood and implemented notions of privacy for far longer than any of the Western societies discussed in the earlier chapters. Ancient Indian texts provide various subtle indications of this in different contexts. From the *Ramayana* it appears that there was an inviolable rule that a female could not be seen by a male stranger. The *Grihya Sutras* spell out, in some detail, the manner in which houses need to be constructed in order to ensure the privacy of their inmates – particularly in the context of maintaining the sanctity of the home while performing religious ceremonies. Kautilya's *Arthashastra* makes reference to the notion of consent by stipulating that no one can enter into another man's house without the permission of the owner.

None of this should come as a surprise to anyone. After all, India has been an advanced, highly civilised society for millennia. It stands to reason that if, when other nation-states evolved from hunter-gatherers into city dwellers they adopted notions of privacy in the process, India too must have done so in similar fashion.

But since we never experienced any of the technology-driven changes that Western societies have gone through, we have never felt the same visceral opposition to its impact in the way that other nations have. Probably the very first time we ever had to think about these issues as a nation was after we had shrugged off colonial rule and were sitting down to write ourselves a Constitution.

When our founding fathers began to draft India's foundational document, one would have thought that personal privacy would have been at the top of their agenda. After all, India had just won independence from the British, who had for a couple of centuries repressively ruled over the nation, denying the 'natives' almost all their personal liberties. The rest of the world had just been through World War II, a period of human history during which people's civil liberties and personal privacy were selectively stripped away on the basis of their race or religious upbringing – and even though that war never spilled on to India's shores, there was an appreciation of its history among the members of Indian society who were responsible for drafting the Constitution. Even though Indians might never have personally experienced the nuances of technologies that deprived us of our privacy, if ever there was a time when privacy ought to have been front and centre in everyone's minds, it was then.

And yet, when the Constitution of India was adopted with much fanfare on 26 January 1950, it contained no mention of the right to privacy.

I have often wondered why this was the case. Was it mere oversight or was there a more deliberate reason? The official transcripts of the Constituent Assembly Debates contain no more than a passing reference to the right to secrecy of correspondence. A speech by Mr R.K. Sidhwa contains the following comment:

> I might also state that the Committee had suggested that the secrecy of correspondence should be guaranteed and that there should be no kind of interception of correspondence, telegrams and telephones, but the main Committee has deleted it. Therefore, it is unfair to say that the Fundamental Rights Committee did not consider this question.[1]

This single throwaway remark, which was clearly part of a longer debate, seemed to confirm my suspicions. It appeared that the framers of our Constitution had considered including a right to privacy in the Constitution but, for some reason that was not entirely clear from the official transcripts, had decided to leave it out of the final draft. If nothing else, this proved that the fact that the Indian Constitution does not mention privacy was not an accident but the outcome of a conscious choice.

I turned to Benegal Shiva Rao's *The Framing of India's Constitution*, easily the most extensive collection of debates and discussions of the various sub-committees that actually worked on the text of the Constitution. Rao had a ringside view of the work of the Constituent Assembly not just because he was a member himself but because he was the brother of B.N. Rau, the person widely recognised as being the true architect of the document and who was responsible for developing many of its more nuanced positions. As a journalist and an academician with these unique qualifications, B. Shiva Rao's perspectives on the framing of the Indian Constitution are invaluable.

The first mention of a right to privacy in his tome was in K.T. Shah's 'Note on the Fundamental Rights' in December 1946, where he discussed the history of fundamental rights around the world and provided a list of what he considered were the essential rights. Included among these was a right to privacy, framed, as is consistent with the thinking of the time, as an aspect of the right to liberty:

> The most important of these relate to the liberty of the person and privacy of the home. No interference of that right can be allowed without due process of law. This is a guarantee against arrest, imprisonment or detention without due process of law, or search warrants of a general character, invasion of the home and the like. Unlike the absolute monarchy of the days gone by, these had been amongst

> the principal grievances of the common people. It is now generally admitted that these are conditions essential and indispensable for living on any decent level of existence.[2]

Early on in its deliberations, the Constituent Assembly established an Advisory Committee to prepare draft articles on fundamental rights and the rights of minorities. This committee had overall supervision over five sub-committees, including the Fundamental Rights Sub-Committee. During various meetings of the sub-committee, distinguished members like K.M. Munshi, Harman Singh and Dr Ambedkar strongly promoted the need to include a right to privacy as one of the fundamental rights.

When Munshi presented the first draft articles on fundamental rights on 17 March 1947, it included Sub-Article (1), which stated that every citizen, within the limits of the law of the Union, should have:

> (e) the right to be informed within twenty-four hours of his deprivation of liberty by what authority and on what grounds he is being so deprived
> (f) the right to the inviolability of his home
> (g) the right to the secrecy of his correspondence
> (h) the right to maintain his person secure by the law of the Union from exploitation in any manner contrary to the law or public morality[3]

Dr B.R. Ambedkar's draft of 24 March 1947 also articulated a similar formulation:

> The right of the people to be secure in their persons, houses, papers and effects against unreasonable searches and seizures shall not be violated and no warrants shall issue

> but upon probable cause, supported by oath of affirmation and particularly describing the place to be searched and the persons or things to be seized.[4]

Based on these individual contributions, the Draft Report of the Sub-Committee on Fundamental Rights dated April 1947 included a specific mention of both the right to secrecy of correspondence as well as the right against unreasonable search and seizure. Article 9(d), which had been adapted from the Weimar Constitution, stated that every citizen should have the right to enjoy secrecy over his correspondence, with the proviso (borrowed from the Indian Post Offices Act) that the legislative could, by law, regulate the interception or detention of articles and messages in the course of transmission in the event of a public emergency or for furthering the interests of public safety and tranquillity. Another clause (borrowed from the Fourth Amendment to the US Constitution) sought to provide all citizens with a fundamental right to secure their person, house, papers and effects against unreasonable search and seizure and stipulated that any such search and seizure could only take place on the basis of a detailed warrant that described probable cause.

While this looks nothing like the broad fundamental right to privacy that one might have expected, as we have seen from our discussions about the development of the law in the US, this sort of formulation was hardly out of place for the times. It recognised that the primary zones within which privacy needed to be safeguarded were the home and personal correspondence. By articulating a fundamental right to the secrecy of correspondence and against unreasonable search and seizure, the framers of the Constitution were attempting to accord constitutional protection to the concept of privacy as it was then understood to exist.

Right from the start, there were strong voices of dissent against the inclusion of these provisions in the Constitution. Many members argued strongly against elevating the right to privacy to the status of a fundamental right. One of the more vocal critics was Alladi Krishnaswamy Ayyar, who voiced his vehement dissent in his comments on the draft:

> In regard to secrecy of correspondence, I raised a point during the discussions that it need not find a place in a chapter on fundamental rights and that it had better be left to the protection afforded by the ordinary law of the land contained in the various enactments...The result of this clause will be that every private correspondence will assume the rank of a State paper or, in the language of Sections 123 and 124, a record relating to the affairs of the State.
>
> A clause like this may checkmate the prosecution in establishing any case of conspiracy or abetment in a criminal case and might defeat every action for civil conspiracy, the plaintiff being helpless to prove the same by placing before the court the correspondence that passed between the parties, which in all these cases would furnish the most material evidence. The opening words of the clause 'public order and morality' would not be of any avail in such cases. On a very careful consideration of the whole subject, I feel that inclusion of such a clause in the chapter on fundamental rights will lead to endless complications and difficulties in the administration of justice. [5]

He was just as critical about Clause 10 relating to unreasonable searches:

> In regard to this subject I pointed out the difference between the conditions obtaining in America at the time when the American Constitution was drafted and the conditions in India obtaining at present after the provisions of the Criminal Procedure Code in this behalf have been in force for nearly a century. The effect of the clause as it is will be to abrogate some of the provisions of the Criminal Procedure Code and to leave it to the Supreme Court in particular cases to decide whether the search is reasonable or unreasonable. While I am averse to re-agitating the matter I think it may not be too late for the committee to consider this particular clause. [6]

But the most influential voice of dissent was that of Benegal Shiva Rao's brother, the constitutional advisor to the Constituent Assembly.

Benegal Narsing Rau was born in Mangalore in 1887 to a family of intellectuals. He graduated with a triple first degree in English, Physics and Sanskrit and went on to study at Trinity College in Cambridge. Thanks to his academic brilliance, he had no trouble getting enrolled in the Indian Civil Service, where he served both as an administrator and as a judge. In the process, he became involved in various administrative projects, including a commission on Hindu law reforms and the Indus Water Commission on river rights.

When the Constituent Assembly was established in 1946, there was little argument when Jawaharlal Nehru recommended that B.N. Rau should be its constitutional advisor. Shortly after he was appointed, B.N. Rau travelled around the world to meet with judges, scholars and authorities on constitutional law to pick their brains on the provisions that the Indian Constitution should contain. In the US he met with Justice Felix Frankfurter, who was at the time engaged in an academic argument with a fellow judge, Justice Black, over the concept of 'due process'. While Justice Black was of the view that the concept needed to be interpreted strictly, Justice Frankfurter, on the other hand, believed that it needed a less stringent approach, requiring no more than the application of principles of 'fairness' or 'ordered liberty'.

There is little doubt that the discussions that Justice Frankfurter had with B.N. Rau were influenced by the public debate he was having at the time. As a result, the latter came away convinced that the inclusion of an obligation to follow the high standards of due process would impose an undue burden on the newly formed Indian judiciary. Immediately following his return to India, B.N. Rau promptly redrafted Article 21 of the Constitution of India to exclude all reference to the concept of due process.

This line of thinking also affected the way he thought about privacy as an offshoot of personal liberty and was probably the reason why he opposed its inclusion as a fundamental right in the Constitution. According to him, such a right could place impediments in the way of law enforcement, particularly given the fact that India was a large country where the administration of criminal justice was bound to be difficult:

> If this means that there is to be no search without a court's warrant, it may seriously affect the powers of investigation of the police. Under the existing law, e.g. Criminal Procedure Code, Section 165, the police have certain important powers. Often, in the course of investigation, a police officer gets information that stolen property has been secreted in a certain place. If he searches it at once, as he can at present,

> there is a chance of his recovering it; but if he has to apply for a court's warrant, giving full details, the delay involved, under Indian conditions of distance and lack of transport in the interior, may be fatal.[7]

His principal objection to the inclusion of privacy as a fundamental right seemed to stem from a concern that allowing for such a right would make the administration of justice in a country as large as India difficult. This seems to indicate a far greater confidence in the capacity of the state to wield its power fairly and without prejudice to the rights of innocent citizens than one would think was warranted.

After several rounds of debates, it was eventually decided that the right to privacy be removed from the chapter dealing with fundamental rights. The final report of the Advisory Committee that was submitted to the Constituent Assembly did not make any mention of the provisions relating to the right to privacy. This was the form of the chapter that was debated and eventually adopted by the Constituent Assembly.

Today we worry far less that criminals will invoke their right to privacy and use it to escape prosecution. We are far more concerned about the ways in which the government can violate the privacy of its citizens if it is not fettered by the constraints that would have been imposed on them had we had a fundamental right to privacy. It seems, with the benefit of hindsight, that while trying to arrive at a balance between the interest of the individual and the objectives of the state, B.N. Rau might have tipped the balance too far in the direction of the state.

B.N. Rau's life till that point in time had been that of a civil servant in the government of British India. His experience of the judicial system was in the administration of justice and not as someone at the receiving end of the injustices that it could perpetrate. The advice he had received while drafting the Indian Constitution had come from an American judge who was similarly presenting a perspective from the other side of the bench – one that ignored the possibility of abuse by the state machinery.

The other opponents of the right to privacy were, similarly, former members of the Indian Civil Service who, until recently, had been responsible for the administration of the country. Their experience of dealing first-hand with the many challenges that came with governance had left them with a strong conviction that it was far more important to vest in the state adequate power to deal with issues of law and order than to equip the individual with a right to protect himself against state overreach. It is not surprising that judges and civil servants aligned themselves with the administrative machinery, trusting with all sincerity that the state would only exercise its powers in the interest of the nation and for the greater good.

Decades later, while I was helping the government draft a privacy legislation for the country, I heard these arguments played back to me again and again. I was told that the threat of terrorism and anti-national aggression demands that investigative agencies be equipped with effective powers of investigation, that the privacy law that I write must contain broad exceptions for law enforcement agencies to ensure that the privacy that we are guaranteeing to the individual does not come in the way of national security and all that the government needs to do in order to secure it.

Time, it seems, does little to change the thrust of these paternalistic arguments. The people in charge of governance, no matter who they are or what sensibilities they might think they bring to the table, always seem to have an unshakeable belief in their own ability to arrive at the appropriate balance between the interests of the state and the privacy of the individual. Those in power will do everything to retain control over their ability to determine how and when national interest will override individual rights. The formulations they come up with are, for that reason, designed to vest in the state machinery's complete authority to decide how that balance is to be arrived at. In the process, individual citizens have little option but to trust that their government will get it right.

That said, in a constitutional democracy, individuals are never completely without recourse. One of the key features of our system of governance is the elaborate checks and balances that have been put in place to ensure that the authority of the state is always kept in check. As a result, even if the state abuses its power to the detriment of individual rights, citizens always have the ability to seek recourse from the independent judiciary against the excesses of the state.

Unfortunately, in the case of the right to privacy, it took nearly six decades for the Indian judiciary to finally come up with a comprehensive formulation of the individual's right to personal privacy.

10

Privacy in the Indian Courts

Just four years after the Constitution of India came into force, the Supreme Court first mentioned the right to privacy in a ruling, even though it was an off-the-cuff remark that had little relevance to the judgment itself. Nevertheless, since the comment was being made by a bench of eight judges, that passing reference was not one that could be easily ignored.

The case in question was M.P. Sharma v. Satish Chandra,[1] and the principle issue was whether a search conducted by the government in the course of its investigations violated an individual's right against self-incrimination. The court had spent some time analysing the historical development of the law of self-incrimination around the world. It noted that in the US, evidence obtained through illegal search and seizure was a violation of the Fourth and Fifth Amendments of the US Constitution. If that principle were to be applied in India, it is possible that the search in question might not have stood up to scrutiny. However, since there is no equivalent of the US Fourth Amendment in the Indian Constitution, the court pointed out:

> ...when the Constitution makers have thought fit not to subject such regulation to Constitutional limitations by recognition of a fundamental right to privacy, analogous to the American Fourth Amendment, we have no justification to import it into a totally different fundamental right.

This throwaway statement, made in passing with no real connection to the substance of the judgment, was the first recorded mention of the right to privacy in the judicial history of post-constitutional India. It was a literal statement reflecting the fact that the Chapter on Fundamental Rights does not explicitly list a fundamental right to privacy. But it would eventually be made to assume much greater significance.

Nine years later, in Kharak Singh v. State of UP,[2] the Supreme Court properly considered the issue in detail for the first time. Kharak Singh had been charged with dacoity and subsequently released for want of evidence. Even though he hadn't been convicted of a crime, he was placed under surveillance as a 'history sheeter', a broad term of reference that allowed the state to keep a watch over him in anticipation that he might commit a crime. As a result, his house was picketed, the police constantly visited his home at night, made enquiries into his habits, the people he associated with and generally made it their business to constantly report and verify his whereabouts. Kharak Singh challenged the constitutionality of the Police Regulations, arguing that the fact that he was being shadowed in this manner was a violation of his personal liberty.

Justice Subba Rao was one of the six judges deciding this case. He wrote a detailed opinion arguing that even though the Constitution doesn't expressly state that the right to privacy is a fundamental right, it is an essential ingredient of the right to personal liberty and that there was nothing more deleterious to a man's physical happiness and health than the calculated interference with his privacy:

> We would, therefore, define the right of personal liberty in Article 21 as a right of an individual to be free from restrictions or encroachments on his person, whether those restrictions or encroachments are directly imposed or indirectly brought about by calculated measures. If so understood, all the acts of surveillance under Regulation 236 infringe the fundamental right of the petitioner under Article 21 of the Constitution.

He had similar things to say about the matter in the context of Article 19(1)(d). He believed that the sort of surveillance being imposed on Kharak Singh was a violation of the right to freedom of movement:

> How could a movement under the scrutinizing gaze of the policemen be described as a free movement? The whole country is his jail. The freedom of movement in clause (d) therefore must be a movement in a free country, i.e., in a country where he can do whatever he likes, speak to whomsoever he wants, meet people of his own choice without any apprehension, subject of course to the law of social control. The petitioner under the shadow of

> surveillance is certainly deprived of this freedom. He can move physically, but he cannot do so freely, for all his activities are watched and noted. The shroud of surveillance cast upon him perforce engender inhibitions in him and he cannot act freely as he would like to do. We would, therefore, hold that the entire Regulation 236 offends also Article 19(1)(d) of the Constitution.

This was a powerful opinion, one that articulated an implicit right to privacy even though no such right was explicitly listed. It was a departure from the originalist approach that the courts at the time were taking in their interpretation of the Constitution – and to that extent was ahead of its time. It was also the first evidence that the line of thinking adopted by the Constituent Assembly in favouring administrative authority over individual liberty was flawed – that the reality of post-constitutional India was not exactly what the founding fathers had imagined it would be.

Unfortunately, Justice Subba Rao was in the minority. The majority toed the line of originalist constitutional interpretation that was in vogue in those early days of Indian constitutional jurisprudence and held that 'the right of privacy is not a guaranteed right under our Constitution and therefore the attempt to ascertain the movements of an individual, which is merely a manner in which privacy is invaded, is not an infringement of a fundamental right guaranteed by Part III'.

Over the next decade, the Supreme Court had no further opportunity to reconsider its views on privacy. As a result, for the first twenty-five years of independent India, the law relating to privacy cleaved closely to the explicit language of the Constitution.

A quarter of a century after the Constitution came into force, the issue of privacy as a fundamental right once again surfaced. In Govind v. State of Madhya Pradesh,[3] the constitutionality of the Madhya Pradesh Police Regulations was challenged on the grounds that domiciliary visits, monitoring of movements and home picketing were a violation of the fundamental right to free movement under Article 19(1)(d) and the right to life and liberty under Article 21. Govind was a convicted criminal who had been marked for surveillance simply because of his criminal record.

The similarities between Kharak Singh and Govind are remarkable. Both judgments dealt with police surveillance and the question of whether this violated the fundamental right to the freedom of movement and the right to life and liberty. What was different was that the Govind case was being decided at a different time.

In the decade that had passed since the Kharak Singh case, two landmark decisions in the United States – Griswold v. Connecticut[4] and Roe v. Wade[5] – had made it clear that privacy could only be denied if there was a compelling state interest to do so. In India, the Gopalan doctrine, which required that each of the fundamental rights be treated as separate and distinct, had been overruled by the decision in Rustom Cavasji Cooper v. Union of India.[6] The judges who decided the Kharak Singh case, with the exception of Justice Subba Rao, had all been following the doctrine that had been laid down in the Gopalan decision, which was why they had used the literal interpretation. Now that the doctrine was no longer binding, the Supreme Court had the freedom to approach the question of privacy from an entirely different perspective.

In deciding Govind v. State of Madhya Pradesh, the Supreme Court analysed the development of law around the world – the two US judgments mentioned above, the wide variety of commentaries and articles on the notion of privacy, and the provisions of the European Convention of Human Rights that expressly articulated a fundamental human right to privacy. Finally, the court spent a long time looking at the decision in the Kharak Singh case, trying to come to terms with its contradictions.

Justice K.K. Mathew, who penned the judgment, was in a quandary. He knew there was no way that his judgment could overrule the Kharak Singh case, since that was a decision of six judges of the Supreme Court and nothing that he said, even if unanimously supported by the other two judges, would overturn the decision of that larger bench. At the same time, he was clearly uncomfortable with the manner in which the majority opinion in the Kharak Singh case had been constructed. It had devoted considerable attention to the elements of privacy, building up the case for an inherent right to personal liberty derived out of the Preamble to the Indian Constitution, but then somehow ended up saying that there was no express right to privacy listed in the Constitution.

So Justice Mathew did the next best thing – he simply ignored the inconsistency. He pointed out that the founding fathers were thoroughly opposed to the Police Raj based on the evidence of the history of the freedom struggle – a statement that is somewhat ironic, given that the very reason why the right to privacy is absent from the fundamental rights is that those drafting the Constitution chose to empower the police and create the very Police Raj that he was railing against. He then obliquely referred to the fact that much had changed since the last judgment on privacy and that the court was obliged to change its interpretation to fit the changing times:

> Time works changes and brings into existence new conditions. Subtler and far-reaching means of invading privacy will make it possible to be heard in the street what is whispered in the closet. Yet, too broad a definition of privacy raises serious questions about the propriety of judicial reliance on a right that is not explicit in the Constitution. Of course, privacy primarily concerns the individual. It therefore relates to and overlaps with the concept of liberty. The most serious advocate of privacy must confess that there are serious problems of defining the essence and scope of the right. Privacy interest in autonomy must also be placed in the context of other rights and values.

Based on this rather tenuous logic, he ruled that mere conviction in a criminal case does not warrant surveillance where there is nothing else that gravely imperils safety. Domiciliary visits by the police can only take place where there is a clear case of danger to security and should not extend to routine follow-up after conviction or at the whim of a police officer. At the same time, he recognised that a right of this magnitude must come with fetters and that as much as there was a need to coax a right to privacy from out of the Chapter on Fundamental Rights, it was necessary to couch this right within appropriate safeguards:

> Assuming that the fundamental rights explicitly guaranteed to a citizen have penumbral zones and that the right to

> privacy is itself a fundamental right, that fundamental right must be subject to restriction on the basis of compelling public interest.

Justice Mathew's opinion made reference to the US decision in Wolf v. Colorado,[7] pointing out that the importance of safeguarding the right to personal privacy against arbitrary intrusion by the police should be as applicable to an Indian home as it is to an American one.

Once the Govind case clearly articulated a right to privacy implicit in the fundamental rights, a number of subsequent cases elaborated on this principle. Over the next forty years, a solid jurisprudence of privacy was built up within the country, Indian courts silently developing case law affirming the right to privacy despite the fact that its Constitution was silent on the principle. These cases touched upon a wide variety of subjects, ranging from the rights of the press to medical privacy. In every one of them, the judges relied on the decisions in these three foundational cases – and particularly on the conclusion arrived at in Govind.

The first time that the judiciary had to strike a balance between an individual's right to personal privacy and the freedom of the press was in connection with a book about the life of a serial murderer.

Auto Shankar was a notorious criminal who had been sentenced to death for six murders. While in prison, he wrote an autobiography and handed it over to his wife with instructions that it be published in a Tamil language magazine. The autobiography laid bare his association with several police officers – many of whom, he alleged, were his partners in crime. The autobiography was due to be published in serial form in the *Nakheeran* magazine and as the date near, the Inspector General of Prisons sent the editors a notice stating that since Auto Shankar had denied writing the autobiography and the prison authorities had no record of having permitted its publication, it was unauthorised. If the editors went ahead with their plans to publish, appropriate legal action would be instituted against them.

The editors took this matter up before the courts in 1995, filing a writ petition restraining the State of Tamil Nadu from interfering with the publication of the autobiography. They alleged that the attempt by the Inspector General of Prisons to prevent the publication was an assault on the freedom of press guaranteed by Article 19(1)(a) of the Constitution of India which entitles them to publish.

The Supreme Court first made it clear[8] that the freedom of press cannot interfere with an individual's right to personal privacy. It said that privacy, which was originally protected under the law of torts, had since acquired constitutional status under Article 21 of the Constitution of India and now guaranteed every citizen, even criminals, the right to be left alone and to safeguard his own privacy as well as that of his family, marriage, procreation, motherhood, child bearing and education. But, having said that, there were exceptions to the rule such as if a person voluntarily thrusts himself into controversy. So long as this publication was based on public records (which would include court records), it cannot be subject to any restrictions. The court also made it clear that public officials do not have a right to privacy with respect to their acts and conduct relevant to the discharge of their official duties. Citizens have a legitimate and substantial interest in the conduct of such persons, and the freedom of press caters to that interest. On that basis, the Supreme Court allowed the autobiography of Auto Shankar to be published irrespective of whether he had consented to it or not.

In the course of the judgment, the court made mention of the one instance in which the right of the press to publish matters in the public record must be curtailed – in relation to cases of rape. The court made it clear that in India, any female subjected to any kind of assault must be spared ruthless publication. It did so to specifically disagree with the principle laid down in Cox Broadcasting Corporation v. Cohn,[9] where the US courts allowed the publication of a rape victim's name on the grounds that it was a part of public record.

This discussion became relevant in a 2013 case[10] brought against the Commissioner of Police, the *Hindustan Times* newspaper and the television news channel Aaj Tak in relation to the disclosure of a First Information Report that contained information about the alleged sexual abuse of a daughter by her own father. *Hindustan Times*, relying on this 'leaked' First Information Report, published the age, locality and class of the girl. The crew of Aaj Tak went a whole lot further, trying to interview the family against its will, all the while telecasting details of their intrusion into the home of the accused, images of the colony in which they lived and personal details of the accused.

The Delhi High Court was harsh in its criticism of the police and said that an integral part of the right to privacy under Article 21 is the maintenance of secrecy of the identity of victims of sexual abuse. A disclosure of this nature was an unacceptable violation of personal privacy. The court also examined the Norms of Journalistic Conduct and pointed out that the press could not reveal the particulars of victims of sexual abuse that could lead to the disclosure of their identity and a resultant breach of their right to privacy. The telecast by Aaj Tak was, in the opinion of the court, a gross violation of the victim's right to privacy.

It is probably interesting to note the defence raised by the media houses in this case, as it is relevant in framing the contours within which the Indian right to privacy operates. They pointed out that a writ petition for violation of fundamental rights is not maintainable against private enterprises since the fundamental rights can only be invoked against the state. This is an important argument, and in order to dismiss it, the Delhi High Court had to refer to the decision of the Supreme Court in Binny Ltd v. Sadasivan, where it was held that fundamental rights are available not only against the state but also against any entity performing a public function for the collective benefit of the public. In the opinion of the court, the press, as the fourth pillar of the state, performed a vital public function and was therefore covered under the ambit of Article 226 of the Constitution of India.

The relationship of matrimony has an unfortunate way of distorting the right to privacy. Courts have often been called upon to decide what rights are available to a husband and wife in the context of their matrimonial relationship.

A young film actress in Andhra Pradesh named Sareetha, who had married Venkata Subbaiah while she was still in high school, separated almost immediately thereafter and lived apart from him. All of a sudden, Venkata Subbaiah filed a petition under Section 9 of the Hindu Marriage Act, 1955, in 1983, seeking restitution of his conjugal rights. Appalled by the thought of being forced into cohabitating with the husband she had parted ways with, Sareetha challenged the constitutional validity of Section 9 on the grounds that any law that allowed her former husband to demand conjugal rights from her violated her right to privacy and dignity under Article 21 of the Constitution of India.

The Andhra Pradesh High Court agreed[11] and held that, since the right to privacy is a part of the right to liberty under Article 21 of the Constitution of India, any law that mandated the restitution of conjugal rights was the grossest form of violation of an individual's right to privacy, denying a woman her free choice to decide when and how her body is to be used. The court relied on the decision in the Planned Parenthood case, where the US Supreme Court had held that the right to privacy belongs to each one of the married couple separately and is not lost by reason of marriage. The court said that no woman who is keeping away from her husband, because of permanent or even temporary estrangement, can be forced, without violating her right to privacy, to bear a child for that man. The court, on that basis, struck down Section 9 as void and unconstitutional. As it happened, this principle was subsequently overruled by the Supreme Court in another case,[12] where the wife wanted to remain in the marriage and was being forced out. The court clarified that the decision in the Sareetha case could not be used to deny a woman the right to remain in a marriage that she very much desired.

Among the many complex issues that need to be addressed in the context of the matrimonial home is the right to privacy in the context of medical health. Given the close personal relationship that a husband and wife share, the ill health of one often affects the other.

When Dharampal tried to divorce his wife Sharda, he needed medical evidence of her insanity to prove his case. Sharda refused to subject herself to a medical examination, and in 2003 Dharampal took the matter all the way up to the Supreme Court. The court had to strike a balance between the wife's right to personal privacy and the fact that if she was indeed insane, her husband would be sentenced to permanent union with a person of unsound mind. The court ultimately concluded[13] that the right to life includes the right to live a healthy life. If Sharda was allowed to successfully avoid a medical examination, it would be impossible to determine whether she was in fact insane. Since there were two competing interests present, there was a need to impose reasonable limitations on the right to personal privacy. The court ordered Sharda to submit herself for medical examination.

In a subsequent case, when Bhabani Prasad of Orissa filed for divorce within three months of his marriage to Nayak, his argument was that their marriage had never been consummated. Nayak countered by filing a complaint before the Orissa State Commission for Women, claiming that she had been tortured by Bhabani Prasad and his family, that she had no source of income and was pregnant. The Commission ordered Bhabani Prasad to provide for maintenance and to ensure that Nayak had a safe delivery. When Bhabani Prasad went to the high court to challenge this order, the court ordered a DNA test to determine the paternity of the child. In 2010, Bhabani Prasad appealed to the Supreme Court of India, arguing that this order violated his privacy. Reflecting on the impact of these new advances in technology, the Supreme Court observed that DNA tests could invade the privacy of the individual in ways that are not only prejudicial to the parties concerned but also the child. Relying on the decision in Sharda's case, the court held[14] that these sorts of tests must be conducted only in cases of eminent need.

In another case, a person who shall remain unnamed – let's call him Mr X – voluntarily donated blood at a hospital, where it was discovered that he was HIV positive. The hospital immediately communicated this information to his fiancée, who promptly called off the wedding. Mr X approached the National Consumer Disputes Redressal Commission on the grounds that information that should have been kept secret had been disclosed. The consumer case was dismissed and Mr X appealed to the Supreme Court of India. While deciding the case in 1999, the court noted that even though a doctor had an obligation to maintain the confidentiality of patient information, there were exceptions to this general rule. Medical information could, for instance, be divulged in cases where it was necessary in order to protect a third party (including a sexual partner) from serious and identifiable risk. The court held[15] that disclosure was justified on the grounds that a woman had the right to be told that the person she was about to marry had a potentially deadly communicable disease.

There has been a conflict between personal privacy and criminal justice since the very beginning of the development of our privacy jurisprudence. Those very early cases were all about excessive surveillance by law enforcement authorities. In time, the questions that had to be addressed by the court came to revolve around the use of new technologies.

When Selvi was accused of kidnapping and murdering Shivakumar, the police had nothing more than circumstantial evidence against him. The police approached the magistrate for permission to subject him to polygraph and brain-mapping tests. He challenged this decision on the grounds that it violated his right to privacy and his right against self-incrimination. The Supreme Court held in 2010[16] that since these tests were drug-induced and entailed physical confinement, it amounted to an intrusion into the accused's mental privacy. Subjecting a person to these techniques without his consent violated the right to privacy. Particularly in the context of criminal proceedings, where the law allows for interference with a convict's physical privacy through permissible arrest, detention and search and seizure, law enforcement agencies cannot use these circumstances as the basis to also compel a person to part with personal knowledge about a relevant fact.

The growth of telecom connectivity helped law enforcement agencies to improve their ability to effectively intercept communications as this became crucial to their ability to detect crimes. When the Central Bureau of Investigation (CBI) issued a report on the 'Tapping of Politicians' Phones', it came to light that there were serious shortcomings in the processes adopted by law enforcement agencies. Records were improperly maintained, interception was carried out beyond the authorised period – in many cases even extended beyond 180 days without the permission of the government. Based on this report, the People's Union of Civil Liberties (PUCL) challenged the constitutional validity of Section 5(2) of the Indian Telegraph Act, 1885, and the process of interception of calls in the country.

The Supreme Court held[17] that phone tapping was a serious invasion of an individual's privacy since telephone conversations are a part of modern life and everyone should have the right to carry on a telephone conversation in the privacy of his or her home or office without interference. Telephone tapping, unless permitted by the procedure established by law, was not only an infringement of Article 21 but also of the right to freedom of speech. This right must be respected and phone tapping should only be resorted to in situations of public emergency and safety, as a last resort when all other methods of acquiring the information had been exhausted. Many of the principles that the court laid down in the PUCL decision correspond to internationally accepted principles of data protection, including the concept of data minimisation that requires those who handle personal data to only collect it for a specified purpose and to keep the collected data in their possession for only as long as is required to fulfil that purpose.

When Reliance Infocom Ltd allowed the interception of politician Amar Singh's telephone conversations in 2011 without giving any thought to whether the orders that had been issued to them by the government were valid or not, the court held[18] that Reliance Infocom was obliged to use its discretion before acting on any and every order of the government. All telecom service providers had to verify the authenticity of requests from the government to see if they were genuine official communications that had been validly issued. Sanctity and regularity in official communication in such matters must be maintained, especially when the service provider was taking the serious step of intercepting a private telephone conversation of its customers.

Banks and financial institutions that are the custodians of personal data have an obligation to act responsibly when dealing with it. Even so, there have been numerous circumstances where banks have either been called upon to release this information or have sought to do so in furtherance of their own objectives. In 2005 in the case of District Registrar and Collector, Hyderabad v. Canara Bank,[19] the court had to decide whether a statute that allowed the Collector to authorise access to documents that had been placed in the custody of a bank was constitutional or not. The Supreme Court held that the right to privacy embodies within it the right to be protected against intrusive observation, and where the right to search and seizure is available to government officials, it should only be invoked to protect necessary state interest.

When Venu defaulted in making payments to the State Bank of India in 2013, the bank threatened to publish his photograph, name and address in leading newspapers if he failed to pay. This name-and-shame approach to loan recovery was challenged on the grounds that it violated his personal privacy and public reputation and, to boot, the Constitution. The High Court of Kerala held[20] that the bank had every right to file a suit for realisation of its dues but could not threaten to publish the names and other details of its defaulters, as no enactment allows them to. Such steps were only permissible under law if Venu was a proclaimed offender and an absconder.

In a subsequent case, when senior lawyer Ram Jethmalani filed a petition asking the government to disclose all documents it had received from the Government of Germany in connection with evidence of unaccounted money stashed in foreign bank accounts, the Government of India resisted this request, arguing that to do so would violate the right to privacy of its citizens. The court agreed, holding[21] that the disclosure of the details of an individual's bank accounts without first establishing whether that person was guilty of wrongdoing was a violation of privacy.

One of the most sensitive areas of personal privacy is sexual preference. There has always been some amount of social stigma associated with the LGBTQ (lesbian, gay, bisexual, transgender, queer) community that has forced them, around the world, to remain on the peripheries of society. In India, in addition to social opprobrium, they have had to deal with the constant threat of criminal prosecution under the provisions of Section 377 of the Indian Penal Code (IPC), which implies that consensual sex between homosexual adults is an unnatural criminal offence.

The Naz Foundation, in what turned out to be a landmark case,[22] challenged the constitutional validity of Section 377 of the IPC, contending that it violated the right to privacy and human dignity of homosexuals. The Delhi High Court agreed, declaring that Section 377 violated the Constitution of India as it criminalises the sexual acts of consenting adults in private. On the issue of privacy, the court held that sexual intimacy is an important facet of human existence. One's sexual preference and personal sexuality is intimate to one's identity and forms a part of the private space. For every individual, whether homosexual or not, the sense of gender and sexual orientation is part of his individuality. The liberty protected under the Constitution allows a homosexual person to enter into relations in his or her own private life. The sphere of privacy allows persons to develop human relations without interference from the community or the state, and homosexuals have as much of a right under the Constitution as heterosexuals.

The court went on to clarify that this provision did not satisfy the test of compelling interest of the state, but, instead, Section 377 had been grossly abused for brutalising homosexuals. The high court also disregarded the argument of morality put forward by the government, stating that the only kind of morality that could pass the test of compelling state interest was constitutional morality, and since the Constitution of India recognises, protects and celebrates diversity, the criminalisation of homosexuality on account of one's sexual orientation would violate constitutional morality.

When the case came before the Supreme Court on appeal, Justice G.S. Singhvi pointed out:

> ...the Division Bench of the High Court overlooked that a minuscule fraction of the country's population constitutes lesbians, gays, bisexuals or transgenders and in the last more than 150 years less than 200 persons have been prosecuted (as per the reported orders) for committing offence under Section 377 IPC and this cannot be made sound basis for declaring that section ultra vires the provisions of Articles 14, 15 and 21 of the Constitution.

He went on to deride the high court judgment by saying:

> In its anxiety to protect the so-called rights of LGBT persons and to declare that Section 377 IPC violates the right to privacy, autonomy and dignity, the High Court has extensively relied upon the judgments of other jurisdictions. Though these judgments shed considerable light on various

> aspects of this right and are informative in relation to the plight of sexual minorities, we feel that they cannot be applied blindfolded for deciding the constitutionality of the law enacted by the Indian Legislature.

With that, in a shocking reversal of what was widely believed to be a progressive judgment, the Supreme Court in 2013 re-instated the criminality of Section 377 of the IPC.

The story of gay rights in India might well have ended there had it not been for the fact that, in a completely different context, the very basis of the fundamental right to privacy was being challenged. In order to decide that case, the largest number of Supreme Court judges ever to gather to deliberate on the issue of privacy was assembled. Even though privacy of the LGBTQ community was not specifically in question in that case, the judges made it a point to specifically call out Justice Singhvi's opinion in the Naz Foundation case, expressing their strong disagreement with the rationale on which he had struck down the judgment of the high court:

> The rights of the lesbian, gay, bisexual and transgender population cannot be construed to be 'so-called rights'. The expression 'so-called' seems to suggest the exercise of a liberty in the garb of a right which is illusory. This is an inappropriate construction of the privacy-based claims of the LGBT population. Their rights are not 'so-called' but are real rights founded on sound constitutional doctrine. They inhere in the right to life. They dwell in privacy and dignity. They constitute the essence of liberty and freedom. Sexual orientation is an essential component of identity.

> Equal protection demands protection of the identity of every individual without discrimination.

This pronouncement was a part of the judgment delivered by the nine-judge bench of the Supreme Court in the 2017 Puttaswamy case, probably the defining privacy decision in the history of Indian constitutional jurisprudence. While the focus of this case was Aadhaar, the court did not pass any judgment as to the validity of the project itself.

In the course of his arguments before the Supreme Court on the constitutional validity of the project, the attorney general of the country had remarked that every single decision in the long line of Indian privacy judgments over the four decades since the Govind case was decided in 1975 had been based on the decisions in the M.P. Sharma and Kharak Singh cases. Both those judgments, he argued, had held that there was no such thing as a fundamental right to privacy in the Indian Constitution and hence, notwithstanding this long jurisprudence of cases, there is no right to privacy in India. It was in response to this provocative statement that a bench of nine judges had been assembled to decide, once and for all, if there was such a thing as a right to privacy in India.

As we have seen, the court has over the years had to deal with a number of cases that required it to determine the privacy implications of using some technology or the other – DNA fingerprinting, modern lie detection techniques, tapping of phones and so on. But never has a technology polarised the country in the way that Aadhaar had. In many ways, Aadhaar is a technology that could only have been created in India. While there is no doubt that a robust identity could have immeasurable benefits in the Indian context, at the same time, it is impossible to accurately quantify the harms that could be caused by deploying something as potent as a frictionless digital identity in the Indian context.

It is fitting that the uncertainty over whether or not we have a fundamental right to privacy would finally be settled in the context of an invention that is uniquely Indian. In order to truly understand the implications of the judgment, we need to spend a little time on Aadhaar itself.

11
Identity and Privacy

Whenever you sign up for a new service in India, you are asked to produce at least two (sometimes more) proofs of identity as part of the registration process. This has become so much part and parcel of modern Indian life that we have come to accept this production of multiple documents as a normal part of the enrolment process. In India, the only way anyone can prove that he is who he says he is, is by cross-referencing multiple sources of identity since no service provider is willing to rely on the veracity of any one of them.

We use different identity documents to interact with different government services in the country. The identity document that probably carries the highest level of assurance is the passport. It is a document that corresponds to international standards and which unequivocally establishes you as a citizen of the country, containing within it details of your age, gender and current residential address. Passports in

India are only issued after an exhaustive verification of your identity, which includes a visit by the local police to your home and confirmation by two references that you actually live at the address you have listed.

Yet, of itself, a passport can't do much more than get you past the immigration counter at the airport. If I offer my passport as a proof of my identity to avail of any services, be they as mundane as obtaining a new telephone connection, I will be asked to produce at least one – sometimes two – other identity documents. How is it that a document, into the production of which the government has invested so much time and effort, is not trusted enough by service providers across the country as to serve as a comprehensive proof of identity?

The same is true of all the other identity documents that the government issues. Tax-paying citizens have to obtain a Permanent Account Number (PAN) from the income tax department that they need to list on their tax returns. Everyone is entitled to a ration card that identifies your eligibility to avail of food rations from public distribution services around the country. Voter identification documents allow you to enter the polling station to cast your vote and, of course, you need a driver's licence to operate a vehicle. Each of these identity cards issued by the government is supposed to be personal to the individual to whom it is issued, offering proof of his or her identity in government records. They are designed to allow that person – and that person alone – to avail of the services offered or regulated by that government department. But when any one of them is individually used as proof of identity, it is never seen as acceptable as adequate proof of personal identity.

The reason for this is our deep history of fundamental mistrust of the government. None of the various processes by which these identity cards were issued is seen to be truly dependable. No single government entity is trusted to have taken the effort to uniquely identify each citizen, and no proof of identity is trusted as being incapable of duplication. As a result, even though the passport department has, for all intents and purposes, done much more to identify a person than is commonly expected in the context of a government identity document, it too is tarred with the same brush, and the identity document it generates is assumed to be as untrustworthy as all the others.

Even though every department of the Indian government has been collecting data from its citizens in various formats for decades, these databases are incapable of talking to each other. Each department collects information in its own unique format and stores it in silos that are not designed to be interoperable. As a result, it is difficult to cross-reference the BPL (below poverty line) database with the LPG (liquid petroleum gas) database, or the telecom subscriber database with the income tax database – despite the fact that the benefit of doing so to check fraud or detect defaulters is obvious.

In India, our identity lies at the intersection of multiple identity documents. We are the composite of the various government services that we avail of. In many ways, this multiplicity of identities is our primary line of defence against identity theft in the absence of a formal law. Since no one has confidence in any one form of identification, in order to impersonate someone's identity, you will have to forge multiple documents. And then, if you want to make the identity theft effective, you will have to repeat that effort with multiple service providers. This is probably why, despite the fact that the government and various private service providers have been collecting data about us for years, there have been no serious incidents of identity fraud. The fact that government databases are maintained in silos, incapable of speaking with each other, and that the proof of identity that they issue is universally mistrusted is what has kept us safe for all these years in the absence of a privacy law.

Most countries around the world have reliable identity systems by which they identify their citizens. This allows government services to be accurately targeted, allowing efficient and effective governance. In India, it is relatively easy for individuals to obtain fake or duplicate identities and as a result government databases are filled with 'ghosts' – persons who are either dead and whose relatives are still claiming benefits using their identity, or who are alive and are intentionally defrauding the system. This absence of reliable identity mechanisms has long been the principal reason why government services rarely reach those they were intended to benefit. The need to improve our systems of identification has been long felt in the higher echelons of power.

In 1993, Chief Election Commissioner T.N. Seshan looked to establish a unique identity for the citizens of India by creating a strong electoral identity card system that they could use to authenticate their identity at polling stations. Following the Kargil war, and in response to various incidents of terrorist incursions into the country, a Group of Ministers, including such political heavyweights as L.K. Advani, George Fernandes and Yashwant Sinha, recommended the creation of a multipurpose National Identity Card as a means to improve the intelligence apparatus. The Citizenship Act was amended to include a new section under which the authority to issue a National Identity Card was established and the Ministry of Home Affairs was made responsible for implementing this project under the National Population Register.

For various reasons, neither of these projects took off. In a country with a long and porous border and an even longer coastline that is impossible to effectively patrol, the voter ID card project struggled because the government was unable to figure out what mechanism it should use to determine who was and who was not an illegal alien. The multi-purpose card was plagued by technological issues and the lack of political consensus. It was apparent that while a unique identity system would benefit vast sections of the society that were currently being denied essential services, the logistical and technological challenges to providing such an identity to every citizen of a country the size of India had defeated the smartest minds that had set out to solve the problem.

When it came to power in 2004, the United Progressive Alliance (UPA) government also recognised that it needed a reliable proof of identity for citizens if it wanted to ensure that subsidies were accurately targeted to the people who needed them the most. In 2006, the government announced that it would implement a unique identity programme for families below the poverty line and placed the National Informatics Centre in charge of implementation. The 'Strategic Vision on the UID Project' recommended that in order to kick-start the process of enrolment, the unique identity should be linked to the only citizen database that was readily available with the government – the electoral rolls.

In much the same way as the endeavours of the preceding Bharatiya Janata Party-led National Democratic Alliance (NDA) government had come a cropper, this project also suffered from the turf wars within the government. Prime Minister Manmohan Singh established an Empowered Group of Ministers to deal with these issues and find a way to bring the various members of his coalition government into alignment. In due course, the Unique Identification Authority of India (UIDAI) was created under the Planning Commission with the objective of collecting resident data in conjunction with the National Population Register.

The multi-purpose national identity card used sixteen fields to uniquely identify a person – name, sex, father's name, mother's name, date of birth, place of birth, marital status, spouse's name, present residential address, permanent residential address, visible identification marks, fingerprint, date of registration, date of issue, date of expiry and a photograph. That said, given that this was going to be a once-in-a-lifetime data-gathering exercise, every department of the government wanted to use the opportunity to collect some information or the other that it needed. As a result, the list was expanded to include blood group, disabilities, religious affiliations, income, etc.

This is not surprising. Data collectors usually prefer to collect more information than less. It is impossible to foresee all the potential future requirements for data, so if you restrict yourself to only collecting data for a specific purpose when other requirements subsequently become evident, you will have to once again undertake the data collection in order to address that purpose. Governments have the tendency to collect as much data as they possibly can, whether or not collection is warranted in the context of the specific purpose for which it is being undertaken. They justify this approach by citing economies of scale that are particularly relevant in the context of a country of our size and population. Collecting identity information from a population the size of India is a mammoth task that is usually only undertaken once a decade. Now that the government was committed to create a unique identity for its citizens, it was understandable that various departments of the government were keen to hitch a ride on that process for their own purposes.

Eventually, it was decided that all that was needed to clearly establish the identity of an individual was the name, age, gender and address along with the email address, mobile number and the father or spouse's name as additional optional fields. These were the data fields that formed the core of the UID dataset and which were eventually ratified by the Demographic Data Standards and Verification Procedure Committee.

Once the UIDAI had been notified as an executive authority that would eventually be granted statutory status, it was left to the authority to decide how the database would actually be built. The government soon realised that the only way a project this large and complex could meet those deadlines was if someone with experience in handling large IT projects in the private sector was given the responsibility of the implementation.

Prime Minister Manmohan Singh agreed and decided that the best man for the job would be Nandan Nilekani.

12
A New Privacy Law

India had built its global reputation as an IT powerhouse on the back of its outsourcing industry. For the longest time, the single biggest issue that every cross-border outsourcing project into India had to deal with was the fact that India had no privacy law. All outsourcing projects involve the transfer of some amount of personal data, and the fact that India had no privacy law was an unacceptable risk to many potential customers as their local laws prohibited them from outsourcing to countries that didn't have adequate data protection regulations.

The Government of India recognised that something needed to be done to address this concern and, in 2008, inserted a new provision, 43A, into the Information Technology Act. This section was designed specifically to address the concerns of the Indian outsourcing industry that was facing pushback from Europe on account of India's lack of a formal privacy law.

The amendment was no more than a single paragraph and it provided general guidance as to how sensitive personal data should be dealt with. It was incomplete in that it left critical terms like 'sensitive personal data and information' undefined and, even though the crux of the new regulation revolved around the security practices and procedures that had to be followed in order to comply with this newly amended provision, no details were provided as to what they were.

For three years after the amendment was introduced, the government remained silent, providing none of the detail that would have made the first general privacy provision to be enacted into law in the country effective. Which is why, when I met Nandan Nilekani for the first time after he had taken up his post as the Chairman of the UIDAI, I was deeply concerned about the privacy implications of the Aadhaar project.

In 2010, I was in the middle of an intense client assignment that required me to commute from Bangalore to Delhi virtually every week. I would catch the early morning flight on Monday to be able to make it to my first meeting in Gurgaon by 10 a.m. and, try as I might to get away as soon as possible, I would inevitably remain there at least till Thursday evening. It was a punishing schedule, but this was the reality of life as a technology lawyer in India. Most of us live and work in Bangalore, which, as the technology capital of the country, is where all the innovation is happening. But since we are the bridge between the regulators and the innovators, we bear the brunt of maintaining that umbilical connection with Delhi.

That morning, as I was boarding my early morning flight, Nandan Nilekani was in the seat next to me. As I crossed him to take my seat, he said with a characteristic twinkle in his eye, 'Welcome to the Monday-to-Friday Delhi brigade.'

I have known Nandan since before he was the CEO of Infosys. Our social circles intersect peripherally, and we had met a few times before that at social events. Given my state of professional anxiety over the privacy implications of Aadhaar, I used the opportunity to discuss my concerns with him in some detail. At the time, based on what little I knew about the project from the press, I was deeply sceptical that a team of technologists would have been able to design a technology solution to achieve a regulatory objective. I was concerned that by following the mantra 'perfect is the enemy of good', they would have sacrificed privacy safeguards at the altar of identity.

Nandan spent the time to take me through the design of the project and the various safeguards that the team had implemented. It was the first chance I had to understand how the entire project had been conceptualised. Over the next few months, I spent a lot more time with Srikanth Nadhumani and others in the technology team, who explained in detail how the system had been designed, the various checks and balances in the workflows and how it innovatively solved the problem of de-duplication at scale. As I began to better understand the architecture of the project, it was evident that considerable effort had been put into getting it to adhere to the principles of data minimisation with appropriate checks and balances baked into the design.

While I came away from those conversations cautiously optimistic, I knew that no technology was ever going to be perfect. Despite everything I was shown, it was possible that a hundred things could go wrong, but at that point in time, at least to my critical eye, many important questions had been asked and answered.

Despite the robust technological design, there were two fundamental shortcomings – both of which had little to do with technology and everything to do with the legal framework within which the project was operating. In the first place, the Aadhaar project itself needed legal backing. Undertaking a project of this magnitude without an enabling statute was asking for trouble. Ideally, that law should have included provisions regulating the manner in which the biometric data being collected should be secured and who would have access to it. Secondly, and to my mind more importantly, rolling out an identity project of this scale in India at a time when we have no privacy law to speak of was a disaster waiting to happen.

The problem, as I explained it to Nandan on that flight to Delhi, was that if the unique identity he was creating was going to be as good and trustworthy as he claimed it was, every service provider was going to want to work Aadhaar authentication into their processes. Even if Aadhaar was designed to be optional, it would quickly become ubiquitous across all databases – government and private. Once that happened and all the databases in the country had been seeded with a common identity number, they would no longer remain the silos that they then were. Aadhaar would serve as that unifying factor getting these databases to communicate with each other.

For years, the lack of a privacy law hadn't bothered us because the databases that contain our personal information are silos, incapable of speaking to one another. Our privacy has been protected over all these years by this plurality of personal information, and the fact that it is impossible to build any sort of personal profile of an Indian is because no common data architecture exists to link our identity. Aadhaar, as I explained to Nandan, had the ability to strip away that protection.

To his credit, he got the point immediately. He understood that no matter how many safeguards he built into the technological framework of Aadhaar, they needed to be backed by a robust legal framework. After all, technologists would have no control over what the government would eventually do with the identity framework once it got its hands on it. In the absence of a legislation that clearly articulated the constraints within which the Aadhaar number had to be used, it was impossible to ignore the possibility that the technology would be misused.

For my part, I could see the benefits that this identity framework would bring to a country where millions were being denied access to the government services to which they were legitimately entitled, simply because they were unable to prove their identity. My concern about the consequences of implementing this identity system in the absence of a full-fledged privacy law was not aimed at bringing the project to a halt. To the contrary, I wanted to ensure that all the good that it could do should be appropriately supported by an appropriate legal framework. It would be a shame, I told Nandan, to have this transformative social benefit scheme derailed simply because the government hadn't taken the trouble to create a privacy law.

Within a few weeks of that chance meeting, Nandan had communicated this concern to the Prime Minister's Office (PMO), which understood the issue and agreed to do something about it. The Department of Personnel and Training (DoPT) was tasked with preparing an approach paper on privacy to determine what should go into a privacy legislation should the country decide to implement one.

A few months later, a joint secretary of the DoPT, Rajeev Kapoor, reached out to me and asked if I would help them draft it.

Rajeev Kapoor was one of those rare civil servants who liked to take the bull by its horns. Data protection was not his area of expertise, but by the time we first met he already had a fair grasp of the broad principles of privacy and fully appreciated how the lack of a privacy law would affect the rollout of the new identity project. As an experienced civil servant, he had a deep personal appreciation of the challenges that the government faced in delivering services, but equally had the ability to conceptualise a future where the benefits of a perfect identity could be turned against the very people who were supposed to benefit from it.

In one of our early meetings, he succinctly articulated the real challenge with introducing a privacy legislation in India. As a civil servant dispensing healthcare benefits to rural India, he told me that he had been required to paint the details of the disease and the treatment provided to the patient on the walls of his or her residence in big bold letters so that everyone in the village could see. This was the government's way of showing everyone in the village that it had done its job of dispensing medicines and, more importantly, that none of the free medical aid that was supposed to be used for treating patients had been diverted to the wrong hands. He knew that this practice violated the personal privacy of the patient and understood why sensitive information should be kept private, but at the same time wanted me to understand the pressure the government was under to demonstrate transparency. In the process, there was bound to be some impact on personal privacy.

This contradiction is quintessentially Indian. Recognising that corruption is endemic, the government had, over the past decade and a half, made a big push for transparency, urging various departments to come up with novel ways in which to demonstrate that the services they were supposed to provide had actually reached the hands of those for whom they were intended. In response, various departments of the government had gone overboard – some painting personal details in public places and others listing this information on their public-facing websites. In both cases, the government seemed genuinely unaware that in doing so it was putting the personal privacy of the people to whom those services had been delivered at risk. For years now, the government had been told to shine light on its activities to demonstrate that it was above board in its dealings. To now be told that there were certain corners towards which that light should not be pointed was going to be deeply disconcerting.

These were the sorts of insights that were critical to the process of preparing a law. Each country needs to arrive at a balance between its many priorities and, as with every other facet of social conduct, privacy is a balance between competing social interests. For as long as humanity has had the concept of personal privacy, it has drawn a line between what must necessarily be made public and what can be kept private. Every now and then, as we have seen, in response to changes in technology, this line gets redrawn and society has to re-adjust to new boundaries.

Given the unique cultural antecedents of our country, it was clear to me that any privacy law we drafted for India was bound to have its own unique flavour. There were many who argued that we had many examples to choose from and that all it would take to create a new privacy law would be to apply one of them with suitable modifications to India. I was reluctant to adopt this approach. Past experience had shown me that law should never be blindly adopted, no matter how similar the circumstances might seem. We needed to create a law that was uniquely responsive to our own cultural requirements.

That said, both Rajeev Kapoor and I agreed on the fact that the first step towards preparing a privacy law for the country would be to understand what went into drafting one. We both saw benefit in learning from global best practices in order to ensure that we didn't end up re-inventing the wheel.

What struck us right at the beginning of this exercise was how many countries around the world already had some sort of a privacy legislation, and that India was possibly the last remaining significant economy without a privacy law. This was a remarkable statistic and a telling one. It was almost unbelievable that the Indian IT industry had managed to succeed under these circumstances, particularly since so many countries in Europe actually prohibited the transfer of personal data to countries that did not have laws that offered at least the same level of personal data protection as Europe.

If you have ever visited an IT facility in India, you will understand why this is the case. The top Indian IT companies have learned to compensate for the lack of statutory protection by implementing state-of-the-art security technologies and business practices. They sign up to binding corporate rules and agree to model contractual clauses that meet the standards of European legislation, ensuring that they adhere to these obligations at all costs. If there is a data breach, it is dealt with quickly and effectively. Every Indian outsourcing company knows that no amount of cost arbitrage will ever compensate for the loss of customer confidence in its privacy practices.

In the two decades that I have been advising the IT industry in India, the instances of data theft have been few and far between – and almost never from any of the top IT companies. In my experience, Indian tech companies have a deep knowledge of what their customers care about when it comes to personal privacy and invest considerable time and effort into ensuring that the services they offer are designed to comply with the laws and regulations that their customers are bound by – even though they are under no compulsion under the law of the land to comply.

Nevertheless, the fact that India was one of the last remaining major economies without a privacy legislation painted the country in a bad light. Over the course of the next few months and years, every time I was called upon to speak with various departments of the government to try and build a broader consensus around the need for the privacy law, I found that mentioning this fact was far more effective than any reasoned argument around civil liberties and personal rights. Even if they didn't really care to engage in a discussion on the nuances of global privacy jurisprudence, I could be sure I would be able to convince any Indian bureaucrat to agree to the need for this law if I told him Pakistan already had one.

As a first step towards the preparation of the approach paper, the DoPT decided to hold a workshop on the legal framework for privacy, data protection and security on 21 July 2010.[1] The DoPT invited a select audience comprising representatives of various departments of the government, including the Ministry of Finance, the Department of Information Technology, NATGRID, the Department of Science and Technology and the Ministry of Home Affairs as well as representatives from civil society and industry. I was asked to make a short presentation to take the assembled bureaucrats through the need for a privacy law and the points that it should contain. The floor was then thrown open for discussion.

At first, it seemed that there was broad acceptance of the need for a privacy law. The Registrar General of India, who was responsible for the census and matters relating to citizenship in the country, pointed out that personal data collected under the Census Act was already required to be kept confidential. However, he did not hesitate to make a reference to the fact that the National Population Register was the 'flagship' government programme responsible for creating a comprehensive biometric-based identity system for the country and that his department was already in the process of making rules under the Citizenship Act. I didn't realise it at the time but this was an indication of the deep schisms that run within the government as to which department should be responsible for the national identity programme – a conflict that would eventually directly challenge the legitimacy of Aadhaar.

As the meeting wore on, more and more evidence of the government's scattered approach to thinking about privacy began to emerge. Various departments were preparing privacy regulations confined to their own narrow sphere of operations without considering the need to address the issue in the broader national context. The Director General of the Department of Information Technology pointed out that the Information Technology Act, 2000, already had a provision relating to data protection and that it should be sufficient to safeguard privacy. The chief legal advisor to the Indian Bank Association indicated that no new law was required for banks since sufficient safeguards were already available within existing law and that the obligation to maintain secrecy with regard to the affairs of the customer was implicit in the contract between the bank and the customer. It seemed that the initial widespread acceptance of the concept of privacy was just a prelude to individual announcements of their own plans to reinforce their own versions of the law.

I worked with the DoPT to prepare the approach paper for legislation on privacy[2] incorporating into it all the feedback from the workshop. My research team reviewed privacy laws from around the world, ranging from the US and UK that had well-developed privacy jurisprudence developed over a long time to that of others like South Africa, which had at the time enacted a new and modern privacy law – and had explicitly included a right to privacy into its Constitution. The purpose behind the exercise was to come up with a definitive list of principles that formed the kernel of all privacy laws around the world so that we had the basic building blocks we needed to shape our own.

The cornerstone of every single piece of legislation that we studied was the principle of consent. Most privacy laws required the data collector to provide notice as to the purpose for which the personal data was being collected and obtain the prior consent of the data subject before collecting it. Laws varied as to the level of detail that was required of the consent being sought and what exactly the notice needed to state. But most of them stipulated that collection had to be limited to only that personal information that was essential for the stated purpose, and that this data, once collected, should only be used for the stated purpose. Data subjects were generally given the right to access their information in order to verify its accuracy and could correct or update the information in case of errors.

I prepared the draft of the approach paper on this basis, setting out brief recommendations for what the proposed privacy law for the country should contain. To the best of my knowledge, this was the first document of its kind in India, and even though this exercise has been repeated at least twice since then, the basic principles that we articulated in that original approach paper on privacy have not shifted materially.

The draft approach paper was circulated among the participants of the workshop and placed on the website of the DoPT for feedback. It received comments from a number of stakeholders – from within the government, civil society and industry. More importantly, now that we had an articulated approach for a privacy law, Rajeev Kapoor began to work the internal machinery of the government, presenting the approach paper to the Committee of Secretaries and recommending that a broad and overarching privacy legislation be enacted in order to ensure that we could establish a unified framework for privacy within which various sectors could draw up their more specific regulations. We had a concern, stemming from our observations at the workshop, that each department was developing regulations specific to itself without formal coordination with each other. We could already see that there were subtle differences in approach that would inevitably result in an uncoordinated patchwork of laws that would pull us in different directions. The Committee of Secretaries agreed with the need for a formal privacy law and tasked the DoPT with the responsibility of preparing a draft. Once he had the go-ahead, Rajeev Kapoor called me to ask if I would help draft it.

We worked on the draft using as reference the material that we had gathered while putting together the approach paper. We pulled from laws around the world appropriate provisions that we believed could be included into an Indian statute. For the most part, the draft law followed the structure that we had set out in the approach paper, covering the broad headings that we had identified. But we also looked for ways in which we could improve on the manner in which these principles were being applied so that we could make compliance less onerous. From my experience with companies that operated in Europe, the procedural requirements under European data protection law were so cumbersome that companies had to invest significant resources in compliance. To protect Indian companies from unnecessarily burdensome obligations we decided to think up novel ways in which to achieve the same results.

One such innovation that ultimately found its way into the draft law was the concept of a National Data Controller Registry. Most privacy laws around the world require data collectors to state upfront the purpose for which they are collecting data and how they intend to use it. The reason for this is to let data subjects know what their data is being used for. That information is usually listed in the privacy policy that the data subject signs up to. Which means that if you want to know what data is being collected from you and how it is being used, your only option is to scroll through every single privacy policy that you have ever signed up to in order to determine what information each of those various data collectors is collecting from you.

In an attempt to simplify this, we decided to create an online portal where the data controllers were obliged to list the data they were collecting and had under their control, and the purposes for which they were intending to use it. It was our intention to make this portal searchable so that important information pertaining to an individual's personal data could be made more readily accessible if required. With that, if anyone was concerned as to how their data was being used, they could access the National Data Controller Registry and search the database for that controller and find out all the different ways in which they were processing personal data.

This was a novel concept and one that we believed was not particularly onerous to implement. The idea was to try and develop a privacy law that was an evolution of what had existed until then, and less than a decade later I would see strains of this idea reflected in the concept of the consent dashboard that the Justice B.N. Srikrishna Committee would suggest in its white paper on the subject.

I soon discovered that the process of enacting a law was much more of a political exercise within the bureaucracy than in Parliament. The privacy law was going to affect multiple departments of the government and so it couldn't be presented to the Committee of Secretaries before it had the broad consensus of all stakeholders. Everyone had their own concerns about what should or should not be included in the draft law and, as a result, the process of finalising the draft took far longer than I had anticipated. While we were still going through these motions, I learned that on 11 April 2011 the Ministry of Communications and Information Technology had issued a set of rules under the Information Technology Act that covered many of the issues that we had intended to address in the draft privacy law.

The rules were called the Information Technology Act (Reasonable Security Practices and Procedures and Sensitive Personal Data or Information) Rules, 2011, and had been enacted by the Ministry under Section 43A of the Information Technology Act, 2000. The first time I read through them, I realised, with a growing sense of irritation, that the contents of the draft rules were remarkably similar to the presentation I had made on the approach paper for the privacy law. Perhaps this was no coincidence since a representative of the IT ministry had been present at the workshop and had been vocal about his disappointment that it was the DoPT, and not them, that had been made responsible for drafting the law.

The Privacy Rules, as I eventually came to call them, contained just eight rules, the first three of which were definitions and other statutory formalities, leaving just five substantive provisions. Within these narrow constraints, the Privacy Rules covered the full range of privacy provisions that one would have ordinarily expected to find in a full-blown statute. It required all bodies corporate that dealt with personal information to publish a privacy policy that stated how they intended to deal with the personal information in their control. There were regulations around the collection of personal information and the requirement for prior written consent. There were restrictions on the collection of information that was not necessary for the stated purpose and the requirement that information was not to be retained for longer than was necessary to achieve the stated purpose. Data subjects were allowed to withdraw consent at any time and the data controller had to take steps to allow the data subject to verify and correct his information as and when required. Finally, in a tip of the hat to the stringent European regulations in this regard, personal data could only be transferred outside India if the entities to which it was being transferred ensured the same level of data protection as was available in India. The Privacy Rules also stipulated that the IS/ISO/IEC 27001 standards were the 'reasonable security practices and procedures' required under Section 43A.

It was remarkable to me that while we were still in the process of enacting a privacy law that would cover all these issues in the level of detail that we believed a piece of legislation of this importance demanded, a department that was fully in the know of the legislative efforts that were under way elsewhere in the government had seen it fit to issue a set of rules that essentially covered the same ground. What was even more galling was the fact that, while the legislative process was fraught with the sorts of checks and balances that is normal in a functioning democracy, the executive seemed to be able to enact a regulation on the same subject matter with hardly any effort at all.

I decided to dig into this in more detail to understand for myself how this had come to pass. The natural place to start was the Information Technology Act itself, the statute under which the rules had been enacted. The power of the government to enact subordinate legislation is circumscribed by principles of administrative law. The legislature cannot abdicate its legislative power by delegating essential legislative functions to the executive and, accordingly, most statutes have a separate section that sets out the specific matters on which the government is empowered to make rules. In the Information Technology Act, this section was Section 87, which listed the rule-making powers of the government. Any rules that were enacted under the IT Act, 2000, had to limit themselves to clarifying statutory provisions that had already been set out in the Act and under all circumstances had to stop short of introducing brand new legislative concepts.

Sub-section 87(2)(ob) is the only provision in the statute that talked about the rules that could be made under Section 43A of the Act. It states that the central government can only make rules relating to 'reasonable security practices and procedures and sensitive personal data or information under Section 43A'. The statute therefore clearly limits the rule-making power of the government to elaborating on what would amount to reasonable security practices and procedures and articulating an appropriate definition of the term 'sensitive personal data or information'.

I now realised why the Privacy Rules had been titled the 'Reasonable Security Practices and Procedures and Sensitive Personal Data or Information Rules of 2011', even though in actual substance they dealt with so much more. The government was only empowered to issue regulations on the definition of the terms 'sensitive personal information' and 'reasonable security practices and procedures' and they clearly did not want to appear to have gone beyond their permitted mandate. But it was obvious, even to a layperson, that in substance the Privacy Rules went much, much further than that. In administrative law terms, the Privacy Rules had clearly been enacted in excess of the rule-making authority granted to the government. If challenged, I had no doubt that the rules would be struck down as being ultra vires the statute under which they were enacted.

To the best of my knowledge, the Privacy Rules have never been challenged. As a result, they operated as the de facto privacy regulation of the country even though they had never gone through a formal legislative process. As a privacy law, these Rules were deeply flawed. They were terse and in their brevity gave rise to considerable internal inconsistency. In some sections they referred to personal information, while in others the focus was on sensitive personal data and information. It was unclear how the law dealt with personal data of citizens of other countries that was being processed in India by outsourcing companies or what the penalty for the violation of any of its provisions was going to be. There was no mention of what must be done when there was a data breach or of who the data subject could approach if his personal data had been misused.

Yet, as this was the closest thing that we had to a privacy law, companies that operated in India were obliged to find a way to comply with its provisions. Over the years I have often been asked to explain how the provisions of this privacy regulation applied to various situations – ranging from what airlines need to do with the data they collect from disabled passengers who are seeking wheelchair assistance, to the personal sensitive data that corporations might accidentally come across in the course of conducting internal investigations on their employees. We have found these questions incredibly difficult to answer with any degree of certainty, given the lack of clarity in the Privacy Rules. And since there is no regulator we can approach for clarifications and no judgments of a court that can be used as guidance, much of what we advise our clients has been based on our own interpretations of the sparse language of the Rules. Given how mature privacy jurisprudence is in most countries around the world, India's half-hearted regulation cuts a sorry figure.

And so we soldiered on, assuming that the Committee of Secretaries was still going to need a draft privacy law, and that when our full-featured law was enacted, it would, in effect, repeat the Privacy Rules.

At around the same time when we were putting the final touches on the draft privacy bill, the country was gripped by revelations that the income tax department had wiretapped the conversations of one of the most prominent PR consultants in the country. Niira Radia was a high-level government lobbyist. In the ordinary course of her business, she had conversations with eminent Indian businessmen like Ratan Tata, Mukesh Ambani and Kalanidhi Maran as well as politicians like A. Raja, the then telecom minister of the country. When the story broke, much of the public furore was in relation to what the tapes seemed to reveal about the extent to which lobbyists could influence both the government and the press on behalf of their corporate masters. On the sidelines of this debate, a discussion around privacy had begun to develop. It became a matter of deep concern in public circles that not only were these conversations being tapped by the tax department, the contents were being allowed to leak to the public and the press. Incensed by this, Ratan Tata filed a case before the Supreme Court invoking his right to privacy.

The indirect consequence of all these events was that when the draft privacy bill was presented to the Committee of Secretaries for its approval, the attorney general returned it with a suggestion that the draft law be amended to include additional provisions addressing the interception of communications, surveillance and direct marketing. To me this was a setback. If we were going to do justice to these three new concepts, we were going to have to undertake a substantial re-write of the law, not to mention a detailed analysis of various existing regulations that already covered, in part, what the attorney general wanted the proposed law to protect. Additionally, it was my belief that concepts like surveillance and interception of communications did not fit into the overall scheme of a data protection legislation which was designed to address issues of personal and sensitive personal data. We already had detailed provisions under the telecom regulations that covered wire-tapping and unsolicited commercial communication, and if we were to incorporate those concepts into our privacy legislation we would have to spend considerable effort aligning these different laws.

That said, we had little latitude to question the wisdom of the Committee of Secretaries or the attorney general, so we attempted to introduce three new sections into the draft. By the time this draft privacy law was finalised, it had passed through the law ministry and had been formatted to include all the bells and whistles that a formal law needed. Rajeev Kapoor had met with most stakeholders within the government and addressed all their concerns. I had personally met with the registrar general of India and the head of NATGRID to offer my assistance in conforming their statutes with the new privacy framework. It appeared that, finally, we had all we needed to enact a privacy law that would provide the kind of substantive privacy protection that the country needed.

But we were yet to suffer our most severe setback. One day, completely out of the blue, I was informed that Rajeev Kapoor had been transferred out of the DoPT and, just like that, the momentum that we had built came to a grinding halt. The draft privacy law would never again have the traction that it enjoyed under his active guidance.

In January 2012, the Planning Commission constituted a group of experts under the chairmanship of Justice A.P. Shah to provide a set recommendations that the government might consider while creating a framework for the privacy law. The group comprised members of various departments of the government, including the director general of the UIDAI; the director general of CERT-In (Indian Computer Emergency Response Team); Rajeev Kapoor, who was at the time still joint secretary with the DoPT, representatives from industry bodies like NASSCOM, civil society and the Planning Commission. The committee analysed the various programmes of the government to assess their impact on privacy and reviewed privacy laws of various countries around the world in order to make specific suggestions that the DoPT could then incorporate into the draft privacy bill. This work was, in many ways, similar to the work we had already done in preparing the approach paper but, by formally constituting a committee under the chairmanship of a retired judge of the high court, the government seemed to be looking to give privacy the level of acceptance that was needed at this stage.

The Shah Committee proposed nine privacy principles on which India's proposed privacy law should be based. The first was notice – all data controllers had to provide easily understandable notice of their information practices before they collected any personal information. This had to include information about what personal information was being collected, for what purpose, how it would be used, whether it would be disclosed to third persons, and what security safeguards had been established by the data controller.

The Shah Committee went on to stipulate that choice and consent were essential elements of the law and data controllers should only collect, process, use or disclose personal information to third parties with the consent of the data subject. Where information is required, by law, to be provided, it should be collected in compliance with the other privacy principles and should be anonymised within a reasonable timeframe if published in public databases. The Committee believed that the data subject should, at any time, have the power to withdraw consent.

It recommended the principle of collection limitation, stressing that data controllers could only collect personal information that was specifically required for the identified purpose and no more. Allied to this was the concept of purpose limitation – restrictions placed on data controllers to only collect and use personal information for the purposes stated in the notice. Once that purpose was achieved, the information had to be destroyed except if there was a statutory data retention obligation.

All data subjects needed the right to access any personal information about them that was being held by a data controller and could request the data controller to correct, amend or delete as appropriate any inaccurate information. The data controller was obliged, on a request by the data subject, to confirm what information it held or controlled about that data subject and obtain for him a copy. Data controllers had to obtain the consent of the data subject before disclosing personal information to anyone else, and all third parties to whom such disclosure was made had to adhere to the privacy principles.

Data controllers also had the obligation to ensure that any personal information under their control was reasonably secured against loss, unauthorised access, destruction, use, processing, storage, modification, de-anonymisation, unauthorised disclosure, either accidental or incidental, and all other reasonably foreseeable risks. They were required to implement practices and policies that were proportional to the scale, scope and sensitivity of the data they were collecting and obliged them to provide information about them in an intelligible form.

Finally, the committee suggested that the data controller must be held accountable for complying with the privacy principles, suggesting that notwithstanding the consent obtained or all the other contractual safeguards that might be in place, the data controller remained responsible for ensuring that in practice the privacy of the individual was secured and protected. This meant they had to put in place mechanisms to implement privacy policies, carry out training and education among their staff and conduct external and internal audits to make sure that the compliance with the privacy principles was not only in name.

These privacy principles took special note of the implications of modern surveillance. They were supposed to be applied across a wide range of sectors and this included the telecommunications sector under which interception and access to communication took place. Once they came into force, any law enforcement officer looking to intercept communications or access private messages would only be able to do so in compliance with these principles. Similarly, the practice of using video and audio recordings for the purpose of surveillance and security would have to be guided by these principles.

The Committee also took the time to talk about the proliferation of personal identifiers like Aadhaar and the fact that these ubiquitous identifiers were causing our siloed databases to converge. It suggested that the national privacy principles be used to determine how information from these converged databases should be used and the manner in which data could be disclosed.

Recognising that it was impossible to have a single law cover all the various sector-specific regulations that would be necessary to roll out a comprehensive privacy regime, the Committee recommended that self-regulating organisations (SRO) in each sector be empowered to create sector-specific policies that drew on the privacy principles to define norms and standards specific to that sector. It saw these SROs functioning in conjunction with the privacy commissioner so that once the standards for a particular industry sector were finalised and submitted to the commissioner, they could be formally approved to operate with binding force in that industry.

For the most part, the recommendations of the A.P. Shah Committee did not diverge too far from the provisions of the draft privacy law in its then current form. Conceptually, there was not much in the nine privacy principles that had not already been covered. Nevertheless, I was called upon by the new joint secretary in the DoPT to refresh the draft and bring it in line with the recommendations of the Committee. Since we were not that far apart, the process of modifying the draft law did not take long, and now that we had the backing of an expert committee that had formally gone into these issues in some detail, I was hopeful that at least this time the draft bill would make it through the legislative process.

Unfortunately, that was not to be. Government in India works in five-year cycles, and towards the tail end of each of these cycles things sort of drift to a standstill. By the time we had completed the modifications and answered all the additional queries that were posed to us by a far more belligerent Ministry of Home Affairs, it was 2013 and elections were around the corner. The government was struggling on many fronts and its interest in pushing the privacy agenda was low. After many follow-ups, I resigned myself to the fact that the draft law I had worked on for three years was going to be consigned to a dusty almirah in one of the rooms of North Block, never to be seen again.

While we were working to prepare the framework for a privacy law, work on ensuring the widespread adoption of Aadhaar was proceeding apace. Despite not having a law to legitimise the use of Aadhaar or to establish the necessary privacy framework within which it needed to operate, the government, with an eye on the approaching elections, began to accelerate its use of the unique identity for social good.

The National Committee on Direct Benefit Transfer (DBT) expanded the use of Aadhaar-based DBT across twenty-seven schemes, resulting in over 5.42 million transactions by the end of 2013. Aadhaar was made a valid document for proof of identity and proof of address that would be accepted when applying for mobile phone connections and LPG cylinders.

It was made mandatory across 291 districts for customers who wanted to access LPG subsidies, eventually covering close to 100 million consumers and facilitating the transfer of more than Rs 33 billion in LPG subsidies. The government authorised its use at voting booths in place of voter ID cards, for the registration of vehicles and obtaining a driver's licence.

The government also began to pave the way for the use of Aadhaar for obtaining passports and as a replacement for the PAN number cited along with income tax filings. The Reserve Bank of India (RBI) allowed banks to accept Aadhaar to open accounts, paving the way for DBT by leveraging Aadhaar-based e-KYC authentication for financial services.

As the pace of deployment of Aadhaar began to ratchet up, consternation began to grow among the opponents of the scheme. For most of them, their initial concern related to the use of biometrics in the design of Aadhaar. They were worried that the government would not be able to adequately prevent it from being abused. As the number of services with which Aadhaar was being linked began to increase, they suddenly realised that the more immediate threat came from the fact that it would soon become ubiquitous and we still did not have a privacy law. They began to mobilise themselves to approach the courts.

On 30 November 2012, the Supreme Court issued a notice in a public interest litigation filed by Justice K.S. Puttaswamy, a retired judge of the Karnataka High Court, challenging the UID scheme, alleging that it had been enacted through a process that was designed to circumvent the legislative process. On 23 September 2013, an interim order was issued in this case, stating that 'no person should suffer for not getting the Aadhaar card in spite of the fact that some authority had issued a circular making it mandatory and when any person applies to get the Aadhaar card voluntarily, it may be checked whether that person is entitled for it under the law and it should not be given to any illegal immigrant.'

Soon after this, India went in for general elections. When the results were announced on 16 May 2014, the BJP and its allies had won a resounding victory and came to power with 336 seats in the Lok Sabha. The incoming prime minister, Narendra Modi, was well known for his use of technology to improve efficiency in governance. During his tenure as the chief minister of Gujarat, he had deployed technology extensively within the state. Recognising the benefits of having a unique identity in order to improve governance, Modi, despite his campaign rhetoric opposing the scheme, decided to adopt Aadhaar wholeheartedly. He called for the revival of the National Identification Authority of India Bill – a piece of legislation that his own party member, Yashwant Sinha, had stalled in Parliament – and enacted it as a Money Bill, avoiding the necessity of formally bringing it before both houses of Parliament.

With the legal validity of Aadhaar now established, he went about incorporating Aadhaar identification into a number of the welfare programmes he had started to roll out. Foremost among these was the Jan Dhan Yojana, the largest DBT scheme that the country had ever attempted. Aadhaar-enabled e-KYC was used for opening new bank accounts and, by 25 January 2015, 115 million new bank accounts had been opened under the Jan Dhan Yojana. As part of this initiative, a collaboration between the UIDAI and the National Payments Corporation of India helped establish a method for offering mobile banking facilities on feature phones. Everyone began to speak of the promise of what soon began to be called the

JAM trinity – Jan Dhan, Aadhaar and Mobile – and how it could be used to provide benefits to the people. The Aadhaar project had greater impetus than ever before and soon crossed the magic number of 1 billion registered residents on its platform.

13

The Puttaswamy Judgment

Civil liberties activists like Usha Ramanathan had always been particularly vocal in their opposition to the project: 'Technology and the machine can, in the land of desperate optimism, seem relatively incorruptible. The potential intrusiveness of technology is shielded by the extent to which the temptations of technology have upended ideas of privacy, confidentiality, personal security and fraud. This seems to have prepared the ground for a technology fix.'[1]

One of the strongest grounds of opposition to the scheme was the concern around mass surveillance. Economist Jean Drèze wrote: 'Most of the "Aadhaar-enabled" databases will be accessible to the government even without invoking the special powers available under the Bill, such as the blanket "national security" clause. It will be child's play for intelligence agencies to track anyone and everyone – where we live, when we move, which events we attend, whom we marry or meet or talk to on the phone. No other country, and certainly no democratic country, has ever held its own citizens hostage to such a powerful infrastructure of surveillance.'[2]

As enrolments reached critical mass, their grouse was not with the original premise of Aadhaar – the certified, verifiable, all-purpose form of identity that everyone seemed to agree would be valuable. What most opponents were agitating against was the relentless acceleration in the deployment of Aadhaar and the manner in which the government was pushing for it to be used as a mandatory identification number across the board.

Once Aadhaar became ubiquitous, the fear was that the government would, with minimal effort, be able to reach across its disparate databases and connect the scattered information about individuals, thereby allowing the government, for the first time, to build a rich, complete picture of their activities. As Malavika Jayaram from the Berkman Klein Centre wrote: 'We should worry about the detailed profiles that it helps create, the complex patterns it reveals when combined with other data, however innocuous, and the social sorting that it enables. Not least because information asymmetries result in the data subject becoming a data object, to be manipulated, misrepresented and policed at will.'[3]

This was the original concern that had prompted the preparation of the draft privacy bill. The very same concerns were articulated in 2010 when I helped draft the approach paper on privacy that was ultimately issued by the DoPT:

> Data privacy and the need to protect personal information is almost never a concern when data is stored in a decentralised manner. Data that is maintained in silos is largely useless outside that silo and consequently has a low likelihood of

> causing any damage. However, all this is likely to change with the implementation of the UID Project. One of the inevitable consequences of the UID Project will be that the UID Number will unify multiple databases. As more and more agencies of the government sign on to the UID Project, the UID Number will become the common thread that links all those databases together. Over time, private enterprise could also adopt the UID Number as an identifier for the purposes of the delivery of their services or even for enrolment as a customer. Once this happens, the separation of data that currently exists between multiple databases will vanish.[4]

The government had missed the opportunity to address these issues at a threshold stage, and pushed ahead with the identity project without first creating the much needed privacy infrastructure within which it was supposed to operate. Now it was the absence of that very framework of privacy legislation that was a threat to the continued existence of Aadhaar itself.

Multiple writ petitions dealing with various issues relating to Aadhaar had been filed before the Supreme Court over the years. Senior Counsel Shyam Divan, appearing for the petitioners, argued that about 90 per cent of the population had Aadhaar cards and, given the serious privacy concerns around the project, the authorities should be injuncted from proceeding further in obtaining biometrics and circulating them to other entities for commercial purposes.

Representing the government's views in the matter was recently appointed advocate general, Mukul Rohatgi. He assured the court that the Union of India did not share any personal information of an Aadhaar card holder with any other person and that Aadhaar was extremely beneficial in that it facilitated various social benefit schemes such as MNREGA (the Mahatma Gandhi National Rural Employment Guarantee Act), public distribution of food and kerosene, and subsidies in the distribution of LPG. He also clarified that Aadhaar enrolment would only take place with the consent of the individual concerned.

On the basis of these assurances, the court permitted the UIDAI to proceed with its operations after giving wide publicity to the fact that it was not mandatory for a citizen to obtain an Aadhaar card and that the production of an Aadhaar card was not a pre-condition for obtaining benefits otherwise due to citizens. Aadhaar was only to be used for the purpose of distributing food grains, cooking fuel and in the LPG distribution scheme. The court made it clear that information obtained by the UIDAI while issuing an Aadhaar card was not to be used for any other purpose.

In the course of his many arguments in the case, Rohatgi adopted the unconventional tactic of challenging the right to privacy itself, calling attention to the fact that between the M.P. Sharma and Kharak Singh cases, the question as to whether in fact there existed a fundamental right to privacy had never been completely settled, and all the subsequent cases had been built on this shaky precedent. Both sides agreed that there was a certain lack of clarity in the jurisprudence and welcomed the opportunity to have the Supreme Court finally settle the legal position. The court seemed to agree that Indian privacy jurisprudence currently stood on shaky foundations and observed:

> …if the observations made in M.P. Sharma and Kharak Singh are to be read literally and accepted as law of this country, the fundamental rights guaranteed under the Constitution of India and more particularly right to liberty under Article 21 would be denuded of vigour and vitality. At the same time, we are also of the opinion that the institutional integrity and judicial discipline require that pronouncement made by larger Benches of this Court cannot be ignored by the smaller Benches without appropriately explaining the reasons for not following the pronouncements made by such larger Benches. With due respect to all the learned Judges who rendered the subsequent judgments, where right to privacy is asserted or referred to their Lordship's concern for the liberty of human beings, we are of the humble opinion that there appears to be certain amount of apparent unresolved contradiction in the law declared by this Court.'[5]

All the parties in question agreed to place the current litigation around the legality of Aadhaar on hold and first settle the question as to whether or not the fundamental right to privacy was part of the Indian constitutional framework. With that, the three judges of the Supreme Court recommended that the matter be heard by a larger bench. On 18 July 2017, a Constitution bench of five judges of the Supreme Court convened to hear the matter. They too quickly came to the conclusion that they would not be able to overrule (or appropriately affirm) the decision of the eight judges who had decided the M.P. Sharma case. Accordingly, it was decided that the matter needed to be adjudicated by a bench comprising nine judges of the Supreme Court.

And so it came to pass that the largest aggregation of Supreme Court judges ever to hear a privacy matter was convened. They listened to arguments from some of the finest legal minds in India – and spent many hours deliberating on the final verdict.

On 24 August 2017, the nine-judge bench delivered a decision that cumulatively ran into 547 pages and contained six separate concurring opinions. It is the most expansive and deeply deliberated discussion on privacy law in the history of Indian jurisprudence, covering a wide range of topics well beyond its original scope. It is in parts historic and poetic, including references from statutes and decisions around the world as well as pronouncements on issues relating to social liberties in the context of modern India that are only tangentially related to privacy. It is well worth it, even for a layperson, to read the original text.

The primary opinion was written by Justice D.Y. Chandrachud. Before getting into the substance of his meticulously structured opinion, he set the stage using a simple definition of privacy that perfectly exemplified the age-old challenge of privacy law – striking a balance between the human need for privacy and the demand that their continued existence in society places on their personal privacy:

> Privacy, in its simplest sense, allows each human being to be left alone in a core which is inviolable. Yet the autonomy of the individual is conditioned by her relationships with the rest of society. Those relationships may and do often pose questions to autonomy and free choice. The overarching presence of state and non-state entities regulates aspects of social existence which bear upon the freedom of the

> individual. The preservation of constitutional liberty is, so to speak, work in progress. Challenges have to be addressed to existing problems. Equally, new challenges have to be dealt with in terms of a constitutional understanding of where liberty places an individual in the context of a social order. The emergence of new challenges is exemplified by this case, where the debate on privacy is being analysed in the context of a global information-based society.

He then went through the entire history of Indian privacy jurisprudence, listing all the cases over the years that had addressed the issue, establishing clearly that the doctrinal foundation of Indian privacy jurisprudence was derived from the trinity of the M.P. Sharma, Kharak Singh and Govind cases. In one way or another, all subsequent cases were based on the decisions in these three, and the uncertainty that currently existed derived from the inconsistencies between those judgments. If we were to have any hope of placing Indian privacy jurisprudence back on firm footing, it would have to be by reconciling these three cases. He then went into each of them in detail.

The M.P. Sharma case, he noted, made but a passing mention of the fact that, in the absence of a right to privacy being specifically set out in the Indian Constitution, a provision like the Fourth Amendment to the US Constitution could not be read into our Constitution. Justice Chandrachud pointed out that this was just an observation that was not relevant to the ultimate decision of the court. What's more, that case did not specifically say that privacy could not be protected under any other provision such as Article 21 or under Article 19. Accordingly, the judgment in the M.P.

Sharma case could be overruled to the extent that it held that there was no fundamental right to privacy.

Turning to the Kharak Singh case, Justice Chandrachud observed that it did not even refer to the M.P. Sharma decision. This was a case that had to decide two issues: (i) the validity of a specific regulation that allowed domiciliary visits at night, and (ii) the validity of the rest of the regulation. In order to hold that domiciliary visits were invalid, the court had relied on the fact that the right to life under Article 21 was an amalgam of the right to life, personal liberty and privacy, and therefore nocturnal domiciliary visits were a violation of the fundamental rights under Article 21. When it came to upholding the validity of the rest of the regulation, the court said that given the absence of a specific right to privacy in the Constitution, the rest of the regulation could not be struck down. The observations in the second part were at variance with those in the first and therein lay the internal inconsistency. Justice Chandrachud held that the latter view had to be an isolated observation that could not co-exist with the essential determination rendered in the first place.

He then turned to the Govind case and noted that even though the judgment referred to the decision in the Kharak Singh case (once again without mentioning M.P. Sharma), it had proceeded on an assumption that there was a right to privacy without getting into the contradictions inherent in the judgment. All subsequent privacy decisions had proceeded on this basis, assuming that the right to privacy emerged from the decisions in the Kharak Singh and/or Govind cases without fully reconciling the inconsistency inherent within the two parts of Kharak Singh. They simply followed the first part of Kharak Singh by ignoring the second – perhaps because none of them had the strength of numbers to overrule the M.P. Sharma and Kharak Singh judgments.

Armed with this reasoning, Justice Chandrachud proceeded to overrule the decisions in M.P. Sharma and Kharak Singh, to the extent that both of them held that the right to privacy was not a right guaranteed by the Indian Constitution. He stated that:

> Privacy is a constitutionally protected right which emerges primarily from the guarantee of life and personal liberty in Article 21 of the Constitution. Elements of privacy also arise in varying contexts from the other facets of freedom and dignity recognised and guaranteed by the fundamental rights contained in Part III.

That said, he recognised that, as with all rights, the right to privacy is not an unfettered or absolute right. To the extent that the notion of a fundamental right to privacy is derived from the right to life and liberty under Article 21 of the Constitution, any invasion of privacy could be justified by a law that stipulates a procedure that is fair, just and reasonable. And so Justice Chandrachud laid down a three-fold test, based on which any law that attempts to encroach upon the right to privacy must be tested:

> (i) legality, which postulates the existence of law;
> (ii) need, defined in terms of a legitimate state aim; and
> (iii) proportionality, which ensures a rational nexus between the objects and the means adopted to achieve them.

This is what the government will have to keep in mind as it rolls out laws that infringe on personal privacy. It is the test that the unique identity project will have to pass if it is to remain valid and, perhaps more importantly, it is what the government will have to keep in mind as it extends the reach and scope of Aadhaar's applications.

In delivering his judgment, Justice Chandrachud carried out an extensive analysis of the concept of privacy, not just in India but in countries around the world – in the UK, the US, South Africa and Europe. He also carried out a detailed analysis of the Constituent Assembly debates in order to understand why the right to privacy was not part of the final draft as well as why the due process clause had been removed – though he came to a slightly different conclusion from the one that I have presented earlier in this book.

Taking advantage of the fact that a Supreme Court bench of this size was unlikely to be convened again any time soon, he addressed many judgments that had, in his opinion, been incorrectly decided in the past. In a remarkable example of the Hindu doctrine of a son's pious obligation to set right the sins of his father, he overruled, with panache, the controversial judgment that his father had delivered during the time of the National Emergency in the ADM Jabalpur[6] case, noting that:

> When histories of nations are written and critiqued, there are judicial decisions at the forefront of liberty. Yet others have to be consigned to the archives, reflective of what was, but should never have been...ADM Jabalpur must be and is accordingly overruled.

He similarly sailed into the judgment of the Supreme Court in the case of Suresh Koushal v. Naz Foundation,[7] where

Justice Singhvi had overturned the Delhi High Court's decision to decriminalise homosexuality on the grounds that the apprehensions of a 'minuscule fraction' of the country's population could not be the basis for declaring that a provision of criminal law was ultra vires the Constitution. Justice Chandrachud held nothing back while condemning the judgment:

> Sexual orientation is an essential attribute of privacy. Discrimination against an individual on the basis of sexual orientation is deeply offensive to the dignity and self-worth of the individual. Equality demands that the sexual orientation of each individual in society must be protected on an even platform.

He also spent some time describing what in his opinion constituted the essential nature of privacy, appropriately placing it in the context of society, just as it had been over the millennia during which the concept evolved side by side with the human race.

> Privacy represents the core of the human personality and recognises the ability of each individual to make choices and to take decisions governing matters intimate and personal. Yet, it is necessary to acknowledge that individuals live in communities and work in communities. Their personalities affect, and in turn are shaped by, their social environment. The individual is not a hermit. The lives of individuals are as much a social phenomenon. In their interactions with others, individuals are constantly engaged in behavioural patterns and in relationships impacting on the rest of society. Equally, the life of the individual is being consistently shaped

> by cultural and social values imbibed from living in the community. This state of flux which represents a constant evolution of individual personhood in the relationship with the rest of society provides the rationale for reserving to the individual a zone of repose.

He then looked at privacy in the context of the modern digital age where data is 'ubiquitous and all encompassing' and technology has made life 'fundamentally interconnected'. He observed that our embrace of technology had ensured that our every transaction left electronic tracks containing information about who we are and about our interests.

> The age of information has resulted in complex issues for informational privacy. These issues arise from the nature of information itself. Information has three facets: it is non-rivalrous, invisible and recombinant. Information is non-rivalrous in the sense that there can be simultaneous users of the good – use of a piece of information by one person does not make it less available to another. Secondly, invasions of data privacy are difficult to detect because they can be invisible. Information can be accessed, stored and disseminated without notice. Its ability to travel at the speed of light enhances the invisibility of access to data, 'information collection can be the swiftest theft of all'. Thirdly, information is recombinant in the sense that data output can be used as an input to generate more data output.

He recognised the many benefits of data-driven decision making, particularly in achieving the objectives of social welfare. He recognised that since the state needs to take appropriate steps to ensure that scarce public resources are not diverted to persons who do not qualify as recipients, modern technologies like data mining could be used to ensure that resources reach their intended beneficiaries. In this context, the use of technology in this manner is not only valid but necessary. Similarly, the prevention and investigation of crimes and the protection of state revenue are also legitimate aims of the state that can and should be supported by deploying technology appropriately.

That said, he pointed out that in the Information Age, the risk to privacy can emanate from non-state actors as well, and for this purpose he instructed the Union government to examine and put in place a robust regime for data protection keeping in mind what had been set out in this judgment.

> The creation of such a regime requires a careful and sensitive balance between individual interests and legitimate concerns of the state. The legitimate aims of the state would include for instance protecting national security, preventing and investigating crime, encouraging innovation and the spread of knowledge, and preventing the dissipation of social welfare benefits. These are matters of policy to be considered by the Union government while designing a carefully structured regime for the protection of the data.

Five other opinions were delivered in this case, all in complete agreement with each other, expounding in greater detail on various other aspects of the right to privacy. Justice Jasti Chelameswar discussed the right to privacy in the context of euthanasia and abortion. Justice Sharad Bobde spoke about our instinctive appreciation of privacy that we exercise when we lock our doors, cover our bodies with clothes and secure our computers and phones with passwords.

Justice Rohinton Fali Nariman presented the history of Indian privacy in terms of what he called the 'Three Great Dissents' – the minority judgements of Justice Fazl Ali in the A.K. Gopalan case, of Justice Subba Rao in the Kharak Singh case and of Justice Hans Raj Khanna in the notorious case of ADM Jabalpur. Each of these three judges had disagreed with the majority decision in these landmark cases and had the courage to speak their views in a written dissent. History would show that these dissenting decisions would eventually be followed by subsequent cases and guide the true path of development of the law.

Justice Nariman also took time to rebut the argument that, since the statute book was already filled with laws that contained privacy provisions, there was no need to separately read into the Constitution a fundamental right to privacy where none exists. He pointed out:

> ...the ruling party can, at will, do away with any or all of the protections contained in the statutes mentioned hereinabove. Fundamental rights, on the other hand, are contained in the Constitution so that there would be rights that the citizens of this country may enjoy despite the governments that they may elect. This is all the more so when a particular fundamental right like privacy of the individual is an 'inalienable' right which inheres in the individual because he is a human being. The recognition of such right in the fundamental rights chapter of the Constitution is only a recognition that such right exists notwithstanding the shifting sands of majority governments. Statutes may protect fundamental

> rights; they may also infringe them. In case existing statute or any statute to be made in the future is infringement of the inalienable right to privacy, this Court then be required to test such statute against such fundamental right and if it is found that there is an infringement of such right, without any countervailing societal or public interest, it would be the duty of this Court to declare such legislation to be void as offending the fundamental right to privacy.

Yet, of all the judges who delivered an opinion, it was Justice Sanjay Kishan Kaul who best used the opportunity to explore the concept of privacy in the context of modern technologies. His judgment is remarkable in its analysis of the impact of technology on privacy.

Modern society, he said, is a form of collective existence that imposes duties and obligations on individuals towards the society of which they are a part. Our 'right to be' has always been circumscribed by societal norms but of late is being further affected by technology. Individuals are at risk that the data that has been collected about them will be used by social networks, search engines, e-mail service providers, messaging applications that already have extensive knowledge of their movements, financial transactions, conversations, health, mental state, interest, travel locations, fares and shopping habits. This sort of knowledge about a person can be used to influence his decision-making processes and shape his behaviour.

He called for the formulation of a data protection law that balanced privacy concerns with legitimate state interests. He made reference to the Justice B.N. Srikrishna Committee that had been appointed to identify key data protection issues in India and recommend methods of addressing them. He hoped, as I did, that this would finally lead to the formulation of a privacy law.

The nine judges of the Supreme Court had convened to redress an ancient anomaly in our jurisprudence. After unequivocally establishing that even though it wasn't written into our Constitution, we did in fact have a fundamental right to privacy, they used the unique opportunity of having nine wise men assembled together in one place to reflect on the challenges that lay before us. They recognised that, in order to benefit from all that the modern data-driven world has to offer, we would need to place some fetters on the right to privacy to which we are entitled. None of their observations were Luddite in character; to the contrary, they seemed to recognise that technology could – and should – be used to improve our lives and benefit the less fortunate in ways that might not be possible without technology.

At the same time, they seemed to be mindful of the fact that all of this is easier said than done. The unintended consequences of blindly adopting modern data techniques could be devastating – to the individual and to society. While predictive algorithms can be remarkably prescient in many of the determinations they make, they can equally be surprisingly blind to nuances of human interactions in a complex society.

It is here that a new balance needs to be struck – one that needs to ensure that the ends of governance and societal benefits are achieved without the human and social costs that would come from thoughtlessly implementing new technologies. We have to learn how to take full advantage of these powerful new technologies without, in the process, eroding our personal privacy.

PRIVACY 3.0

14
Striking a Balance

In India, large sections of the population remain unbanked, ineligible, under traditional banking norms, to access financial products. Opening a bank account itself requires the production of a number of Know Your Customer (KYC) documents that many do not have. On top of that, in order to apply for a loan, applicants need to provide historical evidence of their income and expenditure, usually in the form of bank statements and income tax returns. This automatically disqualifies anyone who does not have a bank account or who has never filed a tax return.

People who operate in the small- and medium-scale sector – and have the greatest need for a loan – usually have the hardest time getting one. There are three main obstacles that come in their way: (i) the general lack of identity documentation, collateral and any sort of formal credit history; (ii) the small ticket size and non-standard nature of their loan requirements; and (iii) the widespread geographic dispersion of the people who operate in the sector, making it economically unviable at current cost levels for traditional financial institutions to service them. At present, the credit portfolios of formal financial institutions have no more than a 15 per cent exposure to the MSME (micro, small and medium enterprises) sector,[1] and the total financing gap for the entire sector is estimated to be about $400 billion.[2]

Once Aadhaar had been deployed widely, the government amended the KYC regulations to allow banks and other financial sector entities to use Aadhaar-based e-KYC to enrol customers. The cost of Aadhaar-based verification is about Rs 60 per person, which, when compared to the nearly Rs 1,000 that it costs to conduct a physical KYC verification, represents a huge saving to the lending industry. It has allowed lenders to bring down their operating expenses down by as much as 60 per cent to 80 per cent and spawned an entirely new digital lending industry that provides loans of anywhere between Rs 25,000 and Rs 500,000. These digital lenders deploy various tools to assess the creditworthiness of the applicants, who normally would not qualify for loans, establishing proxies for trustworthiness, predictability of cash flow and entrepreneurship by cleverly cross-referencing various databases of information that are now capable of being reliably compared since they have all been seeded with a common and reliable identity. The total value of loans disbursed by this new class of lender already exceeds $1 billion.

There are many other examples of financial inclusion that have come about as a result of the new data-driven world we find ourselves in. From these use cases, it is very easy to extrapolate the potential of India's modern financial sector into the future. In time, as more and more people are brought into the formal financial sector by these algorithmic techniques, their financial activities will be monitored with greater granularity, creating more and more accurate records of their financial behaviour. Credit aggregators will be able to track financial performance across the full range of financial service providers and build detailed credit scores for these individuals that will improve their future eligibility for loans. As more data is fed into these financial algorithms, the better will be their ability to develop innovative techniques to map data to outcomes. This in turn will allow even more otherwise ineligible people to enter the formal financial sector, ensuring that these services are administered more efficiently and at an affordable cost. Very soon, this objective, metric-based analysis will completely replace the subjective judgment of the local banker, who approves or denies a loan based solely on his own instinctive reaction to the person in front of him.

But just as these sorts of flow-based lending algorithms can bring people into the financial system, they can just as quickly forever exclude them from it. Proxies are often imperfect, and unless they are constantly evaluated, their output can be inaccurate. This is not easy to do since it usually takes a long period of time for errors in the underlying algorithm to manifest themselves. In situations where there is no active feedback loop – as will be the case where a person is denied a loan that he might never otherwise have been entitled to – it will be much harder to detect the defects in the algorithm. Unless these algorithms are constantly re-tuned to take into account marginal aberrations, it will be a long time before the harm caused by decisions they deliver will become evident.

In a country like India, where a large portion of the population sits at the very periphery of the organised financial sector, if these technologies are imperfectly applied, there is every likelihood that dire consequences will be visited on large sections of the population. From the experience of Western countries, we have seen that once we sacrifice our intuition to these algorithms, we will eventually begin to trust them – even over the evidence of our own eyes. When that happens, people who have been denied financial services by an algorithm could find themselves permanently excluded from the financial system.

We would ordinarily have expected our legal system to have been able to handle these sorts of situations. However, traditional data protection models are singularly ill-equipped to deal with these new technologies. Ever since credit rating agencies started to use consent as the safeguard against data protection violations, it became the default method to protect personal privacy. Data controllers obtain consent from their customers at the very beginning of their relationship with them and proceed to use their data on the basis of the permission they have obtained. Since it is impossible to predict all the uses to which data will be put using these new flow-based technologies, the consent that they obtain is usually framed in the broadest possible terms. Applicants for loans are scarcely in a position to negotiate and find themselves bound by these broad terms going forward. From that point onwards, so long as the data is used for the advertised purpose, the data subject will have no claim against the data controller even if the final outcome is that he is denied a service or, worse, permanently labelled a poor credit risk. Having already consented to have his data processed in this manner, he is in no position to object to the conclusion, now that it has gone against him.

There is clearly a need for a better alternative. The benefits that these new data technologies can unlock are undeniable. However, unless we change the way in which we use them, we could end up doing ourselves more harm than good. If India chooses to use consent to protect itself against privacy violations, it could well forgo the benefits of innovation and, at the same time, find it very hard to protect its people against the harms that could be caused to them by algorithms that have run amok.

There are many other sectors in which we are likely to see the benefits of these new data technologies. In the West, algorithmic solutions have been deployed in areas ranging from criminal sentencing to hiring. There is every reason to think that we will see data being used in a similar manner in India. But there is one specific area that I would like to focus on where I think India, in particular, is likely to receive disproportionate benefits from the application of cognitive technologies – and that is in healthcare.

At present, all medical institutions in India – hospitals, clinics and diagnostic laboratories – operate within silos, using proprietary databases designed solely to serve their own specific requirements. These databases are incapable of interfacing with that of any other stakeholder in the ecosystem, as a result of which it has become impossible to drive synergies in the sector. This is currently the cause of much unnecessary duplication of effort and has inevitably resulted in the loss of vital information along the way.

If we could integrate these different databases, getting them to speak the same language when they interact with each other, we will significantly improve the quality of the data we are working with. This would improve efficiency and lead to significantly better medical outcomes. It would also give us the ability to layer on top of this data relevant non-medical information that might add powerful new nuances to the information.

For instance, by layering geographical data on top of diagnostics statistics, it would be possible to create heat maps of rapidly spreading epidemics like dengue and H1N1 that will allow municipal officials to focus their attention on the heart of those heat maps rather than spreading their efforts across the entire city. By applying contextual analysis to symptomatic information, algorithms can improve our ability to treat diseases like strep throat by correlating symptoms with local demographic information in order to better ascertain how those symptoms should be treated. Using longitudinal time series data correlated with genetic information, we can improve our ability to predict the likelihood of certain diseases occurring in the future and mitigate against it.

At the time of writing, India is on the verge of passing a new electronic health records regulation that will require medical institutions to use electronic health records in the normal course. If this regulation can prescribe a common language that all these various databases need to use to communicate, a central taxonomy of metadata that will allow one database to talk with the other, it will be a powerful first step in allowing data to flow freely across institutions. Given the way in which medical institutions in the Western world have developed their IT systems into huge silos that are completely walled off from each other, if India can actually pull this off, it will be one of the few countries in the world to have successfully done so.

But there are many more significant benefits of using a data-driven approach to diagnosis and treatment. The trouble with diseases is that they can occur in an almost infinite number of combinations. When multiple diseases are simultaneously present within a person's body, the observed symptoms are rarely directly predictive of one clearly identifiable disease. There are many other factors that are unique to each individual – the places they have been, the things they have come in contact with, the specific make-up of their personal microbiome – which makes it even harder to diagnose what specifically ails them. Today, doctors are required to analyse all these multiple data points in the course of one short consultation at a clinic and come up with an accurate diagnosis and a treatment plan. It is to the credit of our doctors and the medical system – and perhaps the inherent resilience of the human body – that this system works and the treatment that we are prescribed usually ends up curing us.

Big data algorithms are particularly well suited to solve these types of problems. It is possible to train computers to identify correlations between symptoms and possible diseases and have them indicate the recommended treatment. This is the fundamental premise behind precision medicine, an approach to the treatment and prevention of disease that analyses individual variations in genes, environment and lifestyle to predict more accurately which treatment and prevention strategies will work for which specific person. It is a significant departure from our current one-size-fits-all approach to healthcare, but it will only be able to work if we have access to vast amounts of data from a large cohort of patients. What often comes in the way of this are the restrictions imposed by privacy law.

Medical data is considered to be among the most sensitive forms of personal data and is accorded the highest level of privacy protection. As a result, creating these large cohorts is almost prohibitively challenging. Not only is it difficult to obtain informed consent – since it is impossible to predict, in advance, the correlations one might find between symptom and diagnosis or treatment – it is often difficult to accurately state the purpose to which this data will be put. If we are to take advantage of the benefits that precision medicine can afford, we will need to re-assess the current model of privacy protection to find new ways in which to safeguard our privacy without sacrificing the benefits of these new technologies.

As India starts to use data to drive its medical decisions, it will have to find a way to ensure that, in the process, the personal privacy of its citizens is adequately protected. It is of paramount importance that the laws within which these data systems operate are equipped to adequately safeguard the interests of the patients.

Currently, consent is the primary legal safeguard used to protect against privacy violation. All transfers of data, interconnections between multiple health information systems, and the use of big data analytics need to pass through the gateway of consent before they can be implemented. However, relying too heavily on consent as a safeguard against privacy violations is, in itself, a concern. When you examine the way it operates, consent neither offers the level of protection that transactions in sensitive personal data deserve nor does it provide the flexibility that these new data technologies need in order to scale and grow.

India is about to plunge head-first into the icy cold reality of data-driven decision making without having had the luxury of first building a culture of privacy. Unlike other countries where data-driven decision making insinuated itself into society over decades, India has no previous experience of dealing with dispassionate algorithms that make decisions based on logical thought processes. We are mentally ill-prepared for the enormity of this change, and we will inevitably struggle when we have to deal with this at scale.

That said, the fact that we got to the party so late puts us in the unique position of being able to learn from the experiences of others and frees us from the path dependence that has dogged them. We have the freedom to think beyond consent as the primary basis for protection and have the unique opportunity to design a legal framework that is more in tune with our needs as a country and our time in history. If India is to make the most of this opportunity, it needs to look at privacy through fresh eyes. Its jurisprudence must unlock the value of data and encourage innovation, but at the same time needs to ensure that no harm accrues to individuals as a result.

We must use the experiences of other countries and the challenges they have faced to guide the choices we make. Of these, the choice we must most strongly interrogate is the continued relevance of consent as the primary line of defence.

15

A New Framework for Privacy

Consent has long been the cornerstone of privacy.

There is something instinctively appealing in the logic that no one should be allowed to use my personal information without my permission. After all, my personal information belongs to me, and even if it is not something that is capable of being physically owned, there is an assumption that what is mine cannot be used by someone else without my consent.

All privacy laws around the world, without exception, have been designed on this basis. Most specify that mere consent is not enough – it must be informed consent. This means that the data subject must be made fully aware of what data is being collected and why, the purpose to which that data is being put and for how long it will be used in that manner. It assumes that, with this information, the data subject will be capable of evaluating for himself whether the use of his personal data by the data controller is likely to affect his privacy or not, and that any permission granted is based on this nuanced understanding of the implications of consent on his personal privacy.

Once the data controller has obtained consent in this manner, it will be free to use that data so long as that use is limited to the stated purpose. If any harm subsequently accrues to the data subject on account of a breach of his personal privacy, the data controller cannot be held liable since there is an implicit assumption that the data subject had been adequately informed of all relevant facts and he must have weighed all the possible consequences before providing his consent.

Consent, therefore, serves two purposes. On the one hand, it gives the data subject autonomy over the use of his personal data, giving him the absolute power to decide whether or not to allow it to be used. On the other, once consent has been properly obtained, it indemnifies the data controller from any violations of privacy or other harm that results from the use of that personal data.

When consent was first used by credit agencies, data processing was still in its infancy. Data collection was done manually, and even after computers got involved, they ran on mainframes and so the data was still stuck within organisational silos. Personal data was collected for a specific purpose and was largely incapable of being put to a different use. Once the data subject knew the purpose for which the data controller intended to use his data, he had all the information he needed to provide informed consent. Each time data had to be put to a new use, it needed to be collected afresh and consent had to be obtained again since the data, once collected, wasn't easily portable.

When personal data was used in this manner, consent served its purpose well. It ensured that the data subject had complete control over his privacy and operated as an effective indemnity mechanism for data controllers. Since it was easy to understand the consequences of disclosing personal data, when consent was provided, it was with full appreciation of the repercussions.

In the nearly half a century that has passed since those early days of data collection, much has changed. Today, data is collected, processed, transferred and consumed in too many ways to comprehensively enumerate. Our social activity is logged, parsed and analysed, our shopping habits observed and personal preferences tracked. Every financial transaction that we undertake is recorded somewhere and often correlated against location, age, time of day and a host of other parameters that we are unaware of. The wearable devices we use to measure personal parameters like fitness and health continuously collect personal data from us, storing this information on servers in the cloud in formats that are easily processed by other service providers who use this information in connection with other data to generate valuable insights about our health and activities. In our daily lives, we constantly engage with sensors, not just those on our fitness trackers, but also those in the smart speakers that listen to everything we say and in our phones that we keep on our person at all times, allowing them to record, along with a multitude of other data, our movements across all three axes.

As a result of this, there is an over-abundance of data being collected from us – data that allows the entities who control it to have a far more detailed profile of us than was ever possible. Today, recruiters can not only learn about our past employment history but can cross-reference our job record with life events using our social media history, providing potential employers with a much more complete picture of us as persons than was ever possible. This has enabled us to find employment that is more holistically suited to us as people rather than just a job that matches our academic credentials.

The hundreds of cookies that lie semi-dormant in our browsers analyse our surfing habits, recording what we look at, the videos we pause over as they auto-play and the conversations we have with our friends on email and instant messenger. They do this to glean information about what we care for most deeply and the products that we most urgently need so that they can then pass this information on via mammoth advertising engines to sellers or service providers who can directly fulfil that need. Which is why we find ourselves being served with listings directly tailored to address our very specific needs with a targeted accuracy that feels eerily like someone was overhearing our conversation.

Data is being called the new oil. And it is proving to be slippery to regulate. The insights it provides are of tremendous value to those who control it and can understand it. Layered together with other information, it is capable of creating detailed profiles about us – finely contoured landscapes that generate accurate and detailed snapshots of us as people, capturing information about our behaviour, the environments we prefer and the ways in which we interact with the world around us. It has flowed into every nook and crevice of our modern existence and influences everything we do. And its outcomes are not entirely benign.

It is remarkable that in all this time and despite these tectonic changes in the power and value of data, the legal construct based upon which it is collected and processed has remained largely unchanged. Organisations that collect personal data still need to first obtain consent and tell us what they are going to use the data for. And so they have us agree to the terms of privacy policies – detailed documents that list all the different types of data that they intend to collect from us and the various uses to which they will be put. They assume that once we have read these terms and provided our consent, they are free to collect data from us and use it for all the purposes stated in the document.

Because the many ways in which data can be used today are varied and complex, the amount of information that tends to be crammed into these privacy policies makes them long and hard to understand. On the occasions that I have read them, they are, even for a lawyer trained to cut through the legalese, dense and complex, making them difficult to grasp. As a result, even though we are supposed to read through the terms and conditions before agreeing to the privacy policy, we rarely do so, clicking 'I Agree' without reading what it is we just agreed to.

It has got to the point where, given the number of services we sign up to, we have begun to suffer from consent fatigue. We think of our acceptance of terms and conditions of service as a formality that we need to complete in order to be able to start using the app we have just downloaded. In many instances, given that these social media services and instant messaging applications have become something of a social necessity, we have no option but to sign up since all our other friends use these services and only interact with each other online. Since so much of modern social interaction takes place through these services, staying offline is akin to self-inflicted social ostracism. For all these reasons, we hardly think twice before we agree to the privacy policy we are presented with, trusting that since the service we are signing up to has so many millions of subscribers across the globe, no harm can come out of agreeing to the terms of its privacy policy.

This raises serious questions as to whether consent is still relevant today. If it has come to a point where almost all of us agree to privacy terms without ever looking past the first screen, surely the consent we are providing is meaningless. And yet, since the way we think about privacy is so inextricably enmeshed with the notion of consent, we continue to act out the charade of accepting privacy policies as if this futile exercise will adequately safeguard our privacy.

I like to try out new technologies, and in the past few years have experimented with a wide range of wearable devices. I have tried a number of step trackers – tiny devices worn on the wrist as well as the more elegant style accessories that can be clipped on to your clothing in various ways. I have experimented with body cameras that record images at fixed intervals, eventually serving up a collage of events and activities that occupied your day. And as they came into fashion, I have experimented with a number of smart watches, enjoying the way they combine a multitude of features and use my phone to upload data into the cloud where further computations could be carried out.

What I particularly enjoyed was the manner in which these wearables have been configured so that their databases could be accessed by other services, allowing me to create as many interesting combinations of services as my imagination allows. For instance, by using the website IFTTT.com,[1] I can configure my phone to analyse how much sleep I had the previous night (using the sleep tracker built into my smart watch) and, if I slept for less than six hours, to remind me that I might want to skip my workout the next morning. This sort of customised workflow is possible because the sleep data that my watch has gathered is shareable with services such as IFTTT that can access and then combine it with messaging services like email or SMS to provide timely and seemingly intelligent interventions.

Even as I marvelled at the useful outcomes that these workflows were capable of producing, I was curious to understand how these interacting databases dealt with privacy. I didn't recall ever consenting to be warned about not going to the gym in the morning, or giving my smart watch consent to share my sleep information with my email provider, but I must have done so if that data was being used to send me an SMS alarm. And so I dug into the privacy policies of each of the different services that had been involved with that sleep reminder workflow I had set up. I realised that through a series of consents provided at different unconnected times, I had authorised a chain of events that no single service provider could have independently conceived of as an intended use of the service. And yet from the way in which each of the individual privacy policies had been put together, the purpose for which consent had been procured seemed to stack up with the final outcome.

Modern databases are designed to be interoperable through Application Programming Interfaces (or APIs) that allow easy access to their datasets. Unlike in the past when personal information was stored in silos – valuable to the data collector and no one else – today, the data from all these sources has been designed specifically to be open and shareable. Its true value lies in being connected with other data and, when combined in these unique and unpredictable ways, it provides insights that no one could have imagined at the time the data was collected. These insights help businesses understand their customers better, allowing them to fine-tune their products and services more narrowly, moving from the current one-size-fits-all business models to making available more personalised service offerings.

Privacy policies have been modified over the years to enable all of this so that our consent to allowing various unnamed third parties to connect to our personal data through these specially designed APIs is taken upfront. As I read through the terms of each of these contracts, it struck me that none of them really owned up to the responsibility of protecting my privacy when they used external data sets to generate the outcomes that they did. Whenever one service was connected to any other external application, responsibility for any resultant breach of privacy was expressly disclaimed. As a result, none of the services that have been daisy-chained together in that long workflow has any responsibility whatsoever for the consequences of its joint actions. Each one hands over responsibility down the chain so that at no point in time can any service provider be responsible for the actions of all of them in concert.

Under these circumstances, even the most seasoned privacy professional will be at a loss to comprehend the privacy implications of these interlinked workflows. It is hard enough to understand the ramifications of any one privacy policy. Evaluating the impact of multiple interconnected datasets is next to impossible.

We are constantly generating data – through our smart devices, from our interactions with those around us and as a by-product of our participation on the internet. Each separate element of data that we generate is, of itself, innocuous and may even be irrelevant from a privacy perspective. But when combined with other similarly individual elements of data, it is capable of being transformed into sensitive personal information. As each of these individual data points is layered one on top of the other, patterns and trends emerge from the stacked data to create profiles of a person or generate patterns that are unique to the individual.

Our online personality exists at the interstices of these various layers of data. Businesses are building increasingly accurate personal profiles of us in order to be able to deliver to us products and services that we like and would appreciate receiving. They have devoted considerable effort to aggregating details of our habits, our likes and dislikes and other distinguishing features that make us who we are. They collect every piece of non-personal information that they can from their interactions with us, operating under the premise that it is better to have more data than less.

When individual elements of non-personal data are combined together, it is possible that, by combining this innocuous data, unique profiles of the individuals they relate to can emerge – patterns that reveal insights from data that was otherwise completely unremarkable. Computers are being designed to process these sorts of datasets, enhancing their ability to build detailed snapshots of us that are unique and deeply sensitive. Machine learning algorithms have been designed to process vast volumes of data and arrive at inferences from their analysis that no human would have come to. As a result, information that was originally non-personal is rapidly being transformed into personal sensitive information. Given the way that privacy laws are designed, as long as no personal information is being collected, there is no legal requirement to seek consent. As a result, there are no legal fetters to the use of this data or the processing of it.

To summarise, as well as it has served us over the years, here are three reasons why consent is no longer a feasible means to safeguard privacy:

1. **Fatigue:** Consent worked as originally conceptualised because there were limited reasons to collect data and few alternative uses to which it could be put. It was relatively easy for a data subject to appreciate the consequences of providing consent. This is no longer the case. Data is collected, processed and used in more ways than we can comprehend. We consent to this extensive data collection by signing standard form contracts that are so complex that it makes them difficult to assess. This, combined with the sheer number of contracts we end up signing, leads to consent fatigue and diminished consent: we end up agreeing to terms and providing consent without actually understanding what we are consenting to.

2. **Interconnection:** Modern databases are designed to be interoperable – to interact with other datasets in new and different ways. This allows us to layer multiple datasets in combinations that generate new insights but which, at the same time, create privacy implications that no one can truly understand. Given how hard it is to understand the implications of agreeing to a single privacy policy, appreciating the consequences of allowing these various different databases to interconnect is beyond the ability of the consent construct.

3. **Transformation:** Machine learning algorithms can take elements of non-personal data and make connections between

them by spotting patterns and building complex personal profiles, transforming them in the process into deeply personal, often sensitive, data. Since there is no need to seek prior consent to collect or process non-personal data, relying exclusively on consent as our only protection against privacy violation is ineffective against the harms that can result from the use of these algorithms.

The world is currently suffering from a deep and pervasive data asymmetry. Data subjects have no idea what is being done to their data, where it is being stored and what processes are being applied to it. All that information lies in the hands of the controllers who not only collect as much data as they can, but process it in so many different ways that it has become impossible for us to truly understand what effect that processing is going to have on us. And still our legal system expects the data subject to be able to determine what needs to be done to safeguard his own privacy.

This hardly seems appropriate. How can a data subject be expected to be held to the consent he has provided when it is impossible for him to fully understand the implications of giving such consent.

We need to find a way to reverse the data asymmetry that currently exists. As much as the data subject needs autonomy over his personal data, it is unreasonable to expect him to understand the implications of providing consent under the current circumstances. Consequently, the data controllers should not be allowed to use the consent obtained from the data subject as a defence against any harm that the data subject suffers on account of the way in which the data is processed.

Data controllers, on the other hand, have complete and unrestricted information about the personal data that is in their control. They also have full knowledge and control of the algorithms they use to process this data and when and under what circumstances they apply them. It is the data controller, therefore, that is best placed to appreciate the impact that collection and processing of data can have on the data subject. Surely it makes sense to shift the responsibility of ensuring the personal privacy away from the data subject and to the data controller.

This is the basic premise behind the new Accountability Model that I have been proposing.[2] It recommends a shift in the focus of privacy regulation from relying on the consent of the data subject to requiring the data controller to be accountable for the manner in which the data is processed. This, I believe, is the appropriate way to restore some of the balance in this data-rich world.

Some of the people I have discussed this model with have expressed misgivings about whether the data controllers – those enormous tech companies through which most of our data is funnelled – will take kindly to being held accountable. It is widely believed that the interests of the data controller are rarely aligned with those of the data subject and that they are driven by purely commercial motivations. By virtue of their organisational design, they have to focus on generating shareholder returns, and it seems futile to expect them to have any concern for protecting the personal privacy of their customers. If they have to choose between improving revenues and protecting the personal privacy of their customers, it seems obvious what they are going to do.

This, it turns out, is a rather uncharitable misconception. Corporations tend to be deeply mindful of the privacy implications of their actions. Data protection has begun to assume such significance the world over that corporations are acutely conscious of reputational impact that a data security breach could have on their business. As a result, companies already go to considerable lengths to avoid getting a reputation that they are intolerant of the privacy concerns of their customers. Most companies, even today, consider themselves responsible for protecting their customers' privacy above and beyond the strict requirements of the law. Holding them legally accountable is unlikely to cause a significant change in their behaviour.

The accountability model will impose a penalty on the data controller in the event a privacy breach occurs. In order for a company to be truly accountable for ensuring privacy, it would need to be liable to suffer the consequences of failing to protect the personal privacy of the individual. Under the EU General Data Protection Regulation, the world's latest data protection legislation, a penalty of as much as 4 per cent of global turnover has been suggested for privacy violations. Something along these lines would have the necessary deterrent value to ensure that data controllers take their responsibilities seriously.

That said, the trouble with introducing heavy penalties into the privacy regime is that it is likely to have the unintended consequence of making data controllers excessively cautious in their approach. While caution is always a good thing when dealing with large volumes of personal data, it could operate as an unnecessary restraint on useful innovation.

The law relating to privacy has always sought to balance competing interests. In the current context, the balance that needs to be achieved is between protecting individual privacy and freeing up data to properly inform our decision making. If the law we propose is only focussed on imposing restrictions on what the data controller can do, it will ensure privacy but will stunt the growth of the data economy. Instead, if we can devise a form of protection that will encourage data controllers to keep innovating with new and useful data-driven business models but at the same time protect privacy, we will have reset the balance.

One way to achieve this is to ensure that the regulatory focus is on remediation rather than on punishment. Machine learning algorithms often have unintended consequences. It is impossible to predict with any level of accuracy what the consequences of applying a particular algorithm or data process is going to be. If the data protection law is going to punish the data controller for every small algorithmic mistake that it makes, it will force data controllers to adopt an excessively defensive approach to data breaches and other privacy violations. This would give them every incentive to keep information of such violations secret rather than inform the community of users of the breach while there is still time to rectify it. This, in turn, will have a chilling effect on innovation as data controllers will choose to develop safer, more conservative technologies as opposed to finding more innovative solutions for fear of punitive consequences.

Instead of punishing data controllers for inadvertent errors in their algorithms or in their data processes, the emphasis should be on encouraging the data controller to remediate in a timely fashion. They should only be penalised if, after they become or are made aware of a privacy violation, they fail to remedy it in time.

This requires a new approach to the regulation of data protection. Data controllers should be given the space to innovate with technologies as it is only out of such innovation that new technologies can develop. At all times, they should be required to design their data systems to enhance the privacy of their customers and to enable data subjects to apply techniques such as de-identification and data minimisation wherever possible, as the less personal data that is out there, the lower is the risk of harm. If in the process they make any mistakes that lead to some form of privacy violations, they should be encouraged, as soon as possible, to rectify these harms. So long as they do so quickly and efficiently, they should not be punished. If on the other hand the harm was caused on account of negligence or a mistake, they should be punished to the full extent of the law.

But even this will not fully address the data asymmetry between the data controller and the data subjects. Shifting the focus to accountability and allowing data controllers flexibility in algorithmic design create an environment within which the data controller is encouraged to remain appropriately responsible for ensuring that its data processes are designed to protect privacy while at the same time being incentivised to continue to innovate. However, even with the best intentions, it is possible that design flaws could slip through. To safeguard against this, the model needs to additionally include a construct by which data processes can be independently verified to assess whether any faulty processes that might have escaped the scrutiny of the data controller are influencing the outcomes.

The easiest way to implement this would be through an audit. Most privacy laws around the work require data controllers to carry out data protection impact assessments (DPIA) every time they roll out a new technology. These assessments are designed to evaluate the harm that is likely to occur from the implementation of a new technology. However, it is usually the data controller itself that is responsible for implementing the DPIA and, given that it is in the interests of the data controller to ensure that the impact assessment is positive, this may not be an entirely effective check. Other privacy laws make the data protection authority responsible for conducting audits. While this achieves the objective of an independent third party audit, I fear that the woeful lack of state capacity in India will make this ineffective in the Indian context.

To address this problem in the Indian context, I have suggested the creation of a new category of intermediaries structurally incentivised to operate in the interests of the data subject. These intermediaries should know and understand technology and also have an appreciation of the impact that these technologies will have on personal privacy. Because of their greater technical expertise, they will be far better equipped than the data subject itself to fully understand the meaning of the specific terms and conditions of the privacy policy as well as the impact of the various algorithms and data processes that they have implemented. By putting themselves in the shoes of the data subject, these learned intermediaries can highlight flaws in the algorithms and processes that would harm the privacy interests of the data subjects.

They will perform the role that financial auditors do today – except that instead of financial affairs, they will audit the data practices of the data controllers. They will publish their findings, pointing out flaws that the data controllers themselves might have overlooked. Applying the principle of remediation, data controllers will be allowed an opportunity to rectify the defects pointed out by the intermediaries without punishment, unless it is shown that they operated with malice aforethought.

In time, I can see these intermediaries begin to rate data controllers for their data practices, offering the equivalent of triple A ratings for data controllers that follow high standards of data protection and lower ratings for those who do not. In time, as awareness of the implications of personal privacy increases, companies will be able to attract customers to their services on the basis of the quality of their data practices. As these intermediaries better educate us of the consequences that bad data practices can have on our personal lives, we will each be able to make more informed choices as to the applications and services that are suitable for us – choosing those that offer the exact balance between privacy and convenience that is appropriate for our specific individual needs.

This is a radical new approach to privacy but one that I believe is a viable alternative to the consent-based approach that we have followed all this while. Consent is stifling innovation in data technologies and creating impediments to the free flow of data. It gives the data subject a feeling of control which is, in practice, completely meaningless, given how little we know about the data that is being collected from us. By creating an entirely new model for data protection based on accountability and audit, I believe we will be able to address, in some measure, the data asymmetry that currently exists.

EPILOGUE
In the Fish Bowl Again

Early man voluntarily placed himself in a fish bowl, his individual survival dependent on harnessing the emergent power of the group. He willingly lived his life in the full gaze of his tribe – agreeing to be watched and likewise to watch over – so that he could throw in his lot with others like him. In this, he was not very different from the animals around him, the sum of his tribe much greater than its constituent parts. He had, like all of nature, discovered the power that lay in being deeply connected with others of his ilk – that came from working together in concert, doing more as a group than was ever possible in isolation.

But man was destined for greater things. He had the intellect to develop technology – the knowledge and skill with which to bend the laws of nature and apply them to his will. He learned superior ways to protect himself and improve his lot, relying more on his tools and science than on his fellow men. In the process, he lost the need to be an organic part of a larger whole – trading that for the ability to keep a part of himself hidden from his fellow men, to display a false facade different from his inner self. He had discovered the power of privacy – the ability to think without fear of rebuke or ridicule and all the benefits that come from it – and he had come to value it. He recognised the role it played in the development of the many facets that make us human and the fact that it needed to be protected.

Privacy is not of nature; it is born of technology and is unique to mankind. But as much as it owes its creation to technology, it is technology that is its biggest nemesis. Every step along its evolutionary road, our notions of privacy have been shaped and formed by advancements in technology. When the printing press democratised knowledge by allowing us to publish the written word, it at the same time allowed private correspondence, which would otherwise have remained confidential between the writer and its recipient, to be duplicated and shared with everyone who cared to read it. When portable cameras democratised photography and made anyone and everyone a photographer, it gave rise to candid photography and yellow journalism. When telecommunication technologies shortened distances between people, they centralised our conversations through pipes, allowing them to be tapped into and intercepted.

Each time a new technology has threatened to strip us of our precious privacy, those who anticipated themselves to be most grievously at risk from this new innovation have demanded that the technology be banned, willing to forgo the potential benefits in order to save us from the risk of suffering harm. Each time, we simply readjusted our jurisprudence to accommodate this new technological advancement and, in hindsight, have been none the worse for it.

We stand once again at the same intersection, witnessing a powerful new technology come into its own. We stand to benefit from the many advantages that it offers, unable, even today, to fully appreciate the ways in which we will be able to leverage it to our favour. At the same time, we have already experienced some of the harms that it can cause and can anticipate that there will doubtless be many others we are yet to see. Like every technology that has come before it, this one too will force us to re-examine the way we think about our privacy and question what we are willing to give up in exchange for all that it has to offer.

And yet, there is something strangely familiar about where we find ourselves today. Networked communications and data technologies have made it possible to once again know as much about each other as we did at a time when there were no secrets between men. Social media has made sharing confidences the norm. It is once again socially acceptable to live our lives in full public view, sharing information, photographs and videos of everything we do with anyone who cares to follow us. Many of us no longer seem to care about maintaining a zone of privacy or a carefully curated facade in front of all save those we trust. To the contrary, we seem to have embraced these new technologies, using them to throw open the doors to our life, allowing all and sundry to enter into areas that even a decade ago were considered intimate.

We seem to be back in that village where we started – where everything we do is public knowledge and nothing can be hidden. We interact with each other in much the same way as our forebears used to, across a common digital hearthstone, swapping stories about our days and sharing our life experiences. We grow suspicious and distrustful of people who are reluctant to share, our technologies providing us the tools with which to 'ping' and 'poke' them when they don't respond. It feels like we have come full circle – like we are right back where we started.

Except that the village is no longer a small geographically bound encampment – but the entire planet. The knowledge we are talking about and have access to is the sum of all the information that we have ever known. We are not part of this egalitarian online society because we voluntarily signed up to the rules of the tribe – but because this is the way things are in our modern networked world. We need a new approach to privacy because our old notions of what must stay private are irrelevant in a world where it is trivial to find out everything about everyone.

There is little we can do to alter this future because it is almost already the present. What we can do is understand it and try to re-write the rules to work within this new context. It is in this way that the future of privacy will be written.

Notes

Prologue

1. Komal Gupta and Suranjana Roy, 'UIDAI temporarily halts Aadhaar payments by Axis Bank, two others', Live Mint, 28 February 2017, http://www.livemint.com/Industry/73F92SKvUKxyngjfx7O0aJ/UIDAI-temporarily-halts-Aadhaar-payments-by-Axis-Bank-two-o.html.
2. 'Hacker shows Bengaluru police how he managed to access Aadhaar data', Huffington Post, 6 August 2017, http://www.huffingtonpost.in/2017/08/06/hacker-shows-bengaluru-police-how-he-managed-to-access-aadhaar-d_a_23067080/

1. Naturally Private?

1. Francis Galton, 'Gregariousness in Cattle and in Men', http://www.galton.org/essays/1870-1879/galton-1871-macmillans-gregariousness-cattle-men.pdf.

2. Ibid.

2. In the Fish Bowl

1. Thomas Gregor, *The Mehinaku: Drama of Life in a Brazillian Indian Village* (Chicago: The University of Chicago Press, 1977).
2. Robert Knox Dentan, *The Semai: A Nonviolent People of Malaya – Case Studies in Cultural Anthropology* (Harcourt College Pub, 1979).

3. What Walls Did

1. The city wall of Uruk was constructed by King Gilgamesh. It is on these walls that he is fabled to have inscribed his deeds that form the substance of the famous epic about his life.
2. Samantha Burke, *Delos: Investigating the Notion of Privacy Within the Ancient Greek Home*, DPhil Thesis submitted to the University of Leicester, 2000.

5. Confidences

1. William Blackstone, *Commentaries on the Laws of England* (1769), p. 223.
2. In Semayne's Case, [77 Eng. Rep. 194 (K.B. 1604)], the court held that it was not a felony for a man to defend his house to death.
3. Pope v. Curll, [(1741) 26 Eng. Rep. 608 (Ch.)] also available at http://www.copyrighthistory.org/cam/pdf/uk_1741a_1.pdf.
4. Ibid
5. Gee v. Pritchard [(1818) 36 Eng. Rep. 670] was a case brought by Mrs Gee against her step son, Rev. Pritchard, who had tried to publish his private correspondence with her. The Court

issued an injunction preventing publication on the basis that Mrs Gee had sufficient property rights in the letters to form the basis for such an injunction.

6. Albert v. Strange, [(1848) 41 Eng. Rep. 1171 (Ch.)], also available at http://www.bailii.org/ew/cases/EWHC/Ch/1849/J20.html.
7. In Ashburton v. Pape (1913) 2 Ch.D. 469 (C.A.) (U.K.) the court held that the duty of confidence could equitably be extended to anyone who was aware of the confidential nature of the communications and after so knowing still discloses it.
8. Saltman Engineering Co. v. Campbell Engineering Co. (1963) 3 All E.R. 413 (1948) (U.K.). In the Saltman case the court held that if a defendant was proved to have used confidential information obtained from a plaintiff, without the consent, express or implied, of the plaintiff, he will be guilty of an infringement of the plaintiff's rights.
9. The case of Argyll v. Argyll [1967 Ch. 302 (U.K.)] was just one part of a long and very public divorce between the Duke of Argyll and his wife, allegedly on the grounds of her serial infidelity. While this particular case was decided in her favour, when the court granted her husband's petition for divorce, the presiding judge remarked that the evidence had established that she was a completely promiscuous woman.
10. In Coco v. Clark 1969 R.P.C. 41 (U.K.) the plaintiff argued that he had designed a moped – the 'Coco' – and supplied Clark with confidential design information with a view to embarking on a joint venture. Clark went ahead and manufactured and sold its Scamp moped, which admittedly shared the same type of piston and carburettor as the 'Coco'.

6. The Right

1. Don R. Pember, *Privacy and the Press: The Law, the Mass Media, and the First Amendment* (Washington: University of Washington Press, 1972), p.7.
2. Neil M. Richards and Daniel Solove, 'Privacy's Other Part, Recovering the Law of Confidentiality', 128, http://ssrn.com/abstract=969495.
3. Robert E. Mensel, 'Kodakers Lying in Wait: Amateur Photography and the Right of Privacy in New York, 1885-1915', *American Quarterly*, Vol. 43, No. 1 (March 1991): pp. 24-25, http://www.jstor.org/stable/2712965.
4. Don R. Pember, *Privacy and the Press: The Law, the Mass Media, and the First Amendment* (Washington: University of Washington Press, 1972), p.7.
5. Warren and Brandeis was renamed Nutter, McLenahan and Fish – that was itself subsequently shortened to just Nutter, a firm that operates to this day in Boston as one of the top technology firms in the area.
6. Harry Kalven, Jr, 'Privacy in Tort Law: Were Warren and Brandeis Wrong?', *Law and Contemporary Problems*, Vol. 31, No. 2 (Spring 1966), pp. 326–27.
7. Elbridge L. Adams, 'The Right of Privacy, and its Relation to the Law of Libel', *American Law Review* (1905): p. 37.
8. Alpheus Thomas Mason, *Brandeis: A Free Man's Life* (William Hein & Co, 1946), p. 70.
9. Martin Green, *The Mount Vernon Street Warrens: A Boston Story, 1860–1910* (New York: Scribner, 1990).
10. Charles E. Colman, 'About Ned', *Harvard Law Review* (2016): p. 128.
11. James Madison, 'Property', http://press-pubs.uchicago.edu/founders/documents/v1ch16s23.html.

12. Thomas M. Cooley, *Treatise on the Law of Torts or the Wrongs Which Arise Independent of Contract* (Chicago: Callaghan, 1879).
13. Thomas M. Cooley, *A Treatise on the Constitutional Limitations*, 5th edition (Lawbook Exchange, 1883), p. 521.
14. Samuel D. Warren & Louis D. Brandeis, 'The Right to Privacy', *Harvard Law Review* (1890).
15. Ibid.
16. Ibid.
17. Ibid.
18. Schuyler v. Curtis [15 N.Y.S. 787 (N.Y. Spec Term 1891)] was literally the first case that recognised a right to privacy – so much so that Judge O'Brien, while granting the injunction, said that 'there is no reported decision which goes to this extent in maintaining the right to privacy, and in that respect this is a novel case'.
19. In Roberson v. Rochester Folding Box Co., [71 N.Y.S. 876 (N.Y. App. Div. 1901)] the judges faced a similar paucity of precedent and had to rely on Schuyler. They recognised the need for a right to privacy along the same lines that Warren and Brandeis had argued.
20. On appeal Roberson v. Rochester Folding Box Co. [64 N.E. 442 (N.Y. 1902)] came before seven judges of the New York Court of Appeals.
21. 50 S.E. 68 (Ga. 1905).
22. 277 U.S. 438 (1928).
23. Katz v. United States, 389 U.S. 347; Watkins v. United States, 354 U.S. 178.
24. 381 U.S. 479.
25. 405 U.S. 438 (1972).
26. 505 U.S. 833 (1992).
27. 539 U.S. 558 (2003).
28. 133 S Ct. 2675 (2013).

29. 135 S. Ct. 2584 (2015).

7. The Currency of Information

1. Daniel J. Solove, 'The Origins and Growth of Information Privacy Law', PLI/PAT, Vol. 748 (2003): p. 29. Available at SSRN: https://ssrn.com/abstract=445181 or http://dx.doi.org/10.2139/ssrn.445181.
2. 96 US 727 (1877).
3. Daniel J. Solove, 'A Brief History of Information Privacy Law' in PROSKAUER ON PRIVACY, PLI (2006). Available at https://scholarship.law.gwu.edu/cgi/viewcontent.cgi?referer=https://duckduckgo.com/&httpsredir=1&article=2076&context=faculty_publications.
4. Ex Parte Brown, 72 Mo. 83, 95 (1880).
5. Rowena Olegario, *A Culture of Credit: Embedding Trust and Transparency in American Business* (Harvard University Press, 2009).
6. 'Verbal Testimony by Michelle Brown', https://archive.is/20120921024215/http://www.privacyrights.org/cases/victim9.htm#selection-125.38-125.110.

9. Early Thoughts on Privacy

1. 'Constituent Assembly of India Debates (Proceedings) – Vol. III', 29 April 1947, http://164.100.47.194/Loksabha/Debates/cadebatefiles/C29041947.html.
2. B. Shiva Rao, ed., *The Framing of India's Constitution* (Universal Law Publishing, 2004).
3. Ibid.
4. Ibid.
5. Ibid.
6. Ibid.

7. Ibid.

10. Privacy in the Indian Courts

1. (1954) SCR 1077.
2. AIR 1963 SC 1295.
3. AIR 1975 SC 1378.
4. 381 US 479 (1965).
5. 410 US 113 (1973).
6. (1970) 1 SCC 248.
7. (1949) 238 US 25.
8. R. Rajagopal v. State of Tamil Nadu, AIR 1995 SC 264.
9. 420 U.S. 469 (1975).
10. ABC v. Commissioner of Police, MANU/DE/0334/2013.
11. T. Sareetha v. T. Venkata Subbaiah, AIR 1983 AP 356.
12. Saroj Rani v. Sudarshan Kumar Chadha, AIR 1984 SC 1562.
13. Sharda v. Dharampal, AIR 2003 SC 3450.
14. Bhabani Prasad v. Orissa State Commission for Women, AIR 2010 SC 2851.
15. AIR 1999 SC 495.
16. AIR 2010 SC 1974.
17. AIR 1997 SC 568.
18. (2011) 7 SCC 69.
19. AIR 2005 SC 186.
20. Venu v. State Bank of India, MANU/KE/0723/2013.
21. 2011 (4) ALLMR (SC) 815.
22. 2010 CriLJ 94.

12. A New Privacy Law

1. 'Approach paper for a legislation on privacy', 18 October 2010, No. 17/1/2010-IR, Government of India, Ministry of Personnel, PG & Pensions, Department of Personnel and Training, http://

www.prsindia.org/theprsblog/wp-content/uploads/2011/06/aproach_paper.pdf.

2. Ibid.

13. The Puttaswamy Judgment

1. Usha Ramanathan, 'The myth of the technology fix', http://www.india-seminar.com/2011/617/617_usha_ramanathan.htm.
2. Jean Drèze, 'The Aadhaar coup', *The Hindu*, 15 March 2016, http://www.thehindu.com/opinion/lead/jean-dreze-on-aadhaar-mass-surveillance-data-collection/article8352912.ece
3. Malavika Jayaram, 'Aadhaar debate: Privacy is not an elitist concern – it's the only way to secure equality', Scroll.in, 15 August 2015, https://scroll.in/article/748043/aadhaar-debate-privacy-is-not-an-elitist-concern-its-the-only-way-to-secure-equality.
4. Approach paper for a legislation on privacy', 18 October 2010, No. 17/1/2010-IR, Government of India, Ministry of Personnel, PG & Pensions, Department of Personnel and Training, http://www.prsindia.org/theprsblog/wp-content/uploads/2011/06/aproach_paper.pdf.
5. The order dated 11 August 2012 in WP (Civil) 494 of 2012 issued by J. Chelameswar, J. Bobde and J. Nagappan.
6. (1976) 2 SCC 521.
7. (2014) 1 SCC 1.

14. Striking a Balance

1. What's happening in the missing middle? Lessons from financing SMEs, World Bank, available at https://openknowledge.worldbank.org/bitstream/handle/10986/26324/113906-WhatsHappeningintheMissingMiddleLessonsinSMEFinancing-29-3-2017-14-20-24.pdf?sequence=1&isAllowed=y.

2. Micro, Small and Medium Enterprise Finance in India – A Research Study on Needs, Gaps and Way Forward (November 2012) by IFC in Partnership with Government of Japan, available at https://www.ifc.org/wps/wcm/connect/4760ee004ec65f44a165bd45b400a808/MSME+Report-03-01-2013.pdf?MOD=AJPERES.

15. A New Framework for Privacy

1. IFTTT stands for 'If This Then That', and is a website that allows people without programming skills to link various online services to produce programmatic results.
2. 'Beyond Consent: A New Paradigm for Data Protection', 20 July 2017, Takshashila Institution, http://takshashila.org.in/takshashila-policy-research/discussion-document-beyond-consent-new-paradigm-data-protection/

Index

Acknowledgements

It feels like I have been writing this book for a long, long time. Much of it grew haphazardly out of the many conversations and arguments I have had over the years around the challenges of developing a privacy regime that is best suited for India. So much so that when I actually started to write the manuscript, it didn't take much time at all.

The genesis of the Accountability Model probably dates back to a seminar I attended at the Radcliffe Institute for Advanced Study at Harvard University, where an eclectic group of medical professionals, economists, private companies and lawyers had been gathered together by the Harvard South Asia Institute to try and come up with a useful regime for exchanging healthcare information in India. Many of the ideas that have found their way into the model grew out of the conversations I had with Adrian Gropper, Saurabh Panjawai, Malavika Jayaram and Satchit Balsari over those two days.

Healthcare privacy remains one of the trickiest and most important problems left to solve.

These ideas were taken further when I was invited to be a part of the Technology Sub-Committee of the RBI Committee on Household Finance. Over the course of our many meetings and discussions, the details of a new model of privacy protection responsive to both the promise of fintech and the special circumstances of our current Indian context were eked out. I am grateful to Tarun Ramadorai for his willingness to consider what is still, to many, a radical construct and for including the model in its entirety in the final report of the committee.

The model needed a home, and the Takshashila Institution warmly embraced it. My original paper was challenged, refined and perfected over many discussions with Nitin Pai, Manasa Venkataraman, Ajay Patri, Pavan Srinath, Pranay Kotasthane, Saurabh Chandra and Madhav Chandavarkar until it finally took shape in a document that has since been widely circulated and extensively cited. I am particularly grateful to Manasa and Ajay for distilling my dense ideas into the more easily understandable format that we finally published as the discussion document titled Beyond Consent.

We road-tested the model over a number of workshops, closed-door conferences and informal discussions, where it was probed, questioned, and – I think – finally accepted by a broad cross-section of people deeply engaged in thinking about privacy in India. I particularly benefited from the many discussions I had with Malavika Raghavan of Dvara, Renuka Sane and Vrinda Bhandari at the National Institute of Public Finance and Policy (NIPFP), Bhairav Acharya at Facebook and Sunil Abraham and Pranesh Prakash at the Centre for Internet and Society (CIS). If there is a magic circle of privacy thought leadership in India, this is them.

My initial attempts at drafting a privacy law for the country would not have been possible had it not been for the efforts of Nandan Nilekani, who, even though he is not given enough credit for it, had, as early as 2010, understood the need for a privacy framework within which Aadhaar had to operate. The fact that we do not still have a law is not for want of trying on his part. Much of my understanding of the technical design and architecture of Aadhaar – and consequently the privacy features incorporated by design into its architecture – came from many extended discussions with Srikant Nadhumani and later Pramod Varma and Sanjay Jain. At the time, the burden of conceptualising the regulatory framework rested on the shoulders of Deepika Mogilishetty, with whom I have had numerous long and involved discussions on these issues. Little did we know that five years later the concepts we discussed would be debated threadbare in the highest court of the land by some of the finest legal minds in the country.

When I was called on to help draft the privacy law, I worked with many fine civil servants in various departments of the government. Unlike the unfortunate stereotype that dogs them, these were, to a man, motivated IAS officers who refused to let their unfamiliarity with the nuances of privacy law come in the way of producing a robust draft law. Of particular note was Rajeev Kapoor, who readily admitted at our first meeting that this was his first brush with the concepts of data protection, but soon became a master of its nuances and was singularly responsible for the original draft. K.P. Krishnan, even though he never had direct oversight on issues of privacy, has always been generous with his time and a guiding influence in all my endeavours in the area of policy.

He may not recall it, but he was a quiet comfort during the many days I spent at North Block working on the privacy law.

My current understanding of privacy law would have been incomplete without a global perspective. To the many lawyers and practitioners around the world whom I have worked with over the years, I owe a debt of gratitude. They are too numerous to individually name but I must especially call out the participants of the Cambridge Forum's annual roundtable on privacy who every year engage in some of the most balanced and thoughtful discussions on the subject. As meagre as our country's privacy laws are, our geopolitical importance on the world stage gave me a seat on that table and helped me understand how the rest of the world thinks of privacy.

I have had the opportunity to work on privacy law issues for nearly two decades – well before most people in India engaged with them. For this I am grateful to Trilegal for giving me the freedom and the space to dabble in an area of law that did not exist when I first started working on it. It is this sort of long-term investment and enterprise that has taken Trilegal from a first-generation legal start-up to the highest echelons of the Indian legal industry. Of this I am immeasurably proud. I am particularly grateful to my five closest friends – the partners with whom I started this journey close to two decades ago: Prem Ayappa, Anand Prasad, Akshay Jaitly, Karan Singh and Sridhar Gorthi. Their generosity of spirit, camaraderie and shared vision in the future of the legal industry in India are the stuff of legend.

Trilegal's technology practice is second to none, and the many young lawyers I have worked with over the years are without a doubt some of the brightest and most hard-working technology lawyers in the country. They know that being in this team requires them to willingly engage deeply with the technology they advise on so that they can understand how it works before applying the law to clients queries. It is through discussions with these sharp young minds that many of my theories on privacy were tested, shaped and hardened. I am, in particular, grateful to Esha Goel, who was just a young intern from National Law School when I asked her to find out why we didn't have a fundamental right to privacy in our Constitution. Much of the material she unearthed has made its way into this pages.

My weekly column in *Mint* – 'Ex Machina' – has been the playground where I have experimented with many of the ideas that make up this book. I never thought I would be able to generate 800 words on a new topic every week. Nearly 100 articles later, it just goes to show that you never know what you are capable of until you try. Sukumar Ranganathan took a leap of faith by inviting me to be a member of the exclusive club of *Mint* op-ed writers. Each week I am excruciatingly aware of the high standards I have to live up to.

As a practising lawyer, writing is my daily job. However, writing a book was like nothing I had done before. I am lucky to have had a lot of assistance along the way. Halfway through the first draft of this book, my dear friend Lavanya Sankaran – without question the most accomplished writer I know – grabbed me by the collar and sat me down for a masterclass in writing. Those three hours literally upended the way I thought about the art of authorship. This book is immeasurably better for that timely intervention, but I know I still have a long way to go.

I would never have started writing this book had it not been for my dear friend and literary agent Jayapriya Vasudevan, who has been badgering me to write a book – any book – for fifteen years. I knew I would one day, but had it not been for her gentle and constant persuasion, this might never have come to be. My editor at HarperCollins, Siddhesh

Inamdar, who fought for this book and worked against time to get it out in record time, gave it its final shape. It was evident from his insightful edits that he had worked hard to educate himself on an esoteric subject.

My family has been a constant source of inspiration and support. My father wrote the speech I gave at the very first debate I ever participated in at school, and even though I delivered it like a piece of elocution, it marked the start of a long career of debating and reasoned thinking. My mother, who has ever since borne the brunt of my constant argumentativeness, always told me that no matter what I chose to be I had to become the best in my field. It's crowded out here in the technology space, but I hope this book is a step in the right direction.

My son Dhruv, who over the past year became an integral part of my morning writing schedule, is a constant reminder of how important it is for us to do right by the digital first generation. As I observe how intuitively he engages with technology, I feel compelled to re-double my efforts to make the online space safer and more productive. My lovely wife Ahalya, who has been with me through the entire privacy journey described in this book, has been my constant support though this year of writing, putting up with a distant husband, even though, when compared with the work she is doing documenting the neglected textile history of our country, the contribution of this book pales in significance.

It would be completely out of character for me to end without acknowledging Coco and Zoey, without whom this book would never have been completed. Every writer will tell you that it is important to take regular breaks so that you can keep the juices flowing. For me, that regular reminder was a wet nose lodged under my writing arm, demanding to be petted.

About the Author

Rahul Matthan is a partner with the law firm Trilegal and heads its TMT (technology, media and telecommunications) practice.

He has been working on issues relating to technology law for over two decades and has been involved in a number of policy initiatives at the intersection of law, society and technology. His active involvement in the privacy policy space has given him a unique view into how these issues evolved in the country. He was involved in some of the early drafting of a privacy law that, for various reasons, eventually did not get enacted. He has since served as a member of the Technology Sub-Committee of the Reserve Bank of India's Committee on Household Finance, where he authored the section on privacy. More recently, his Discussion Document entitled 'Beyond Consent – A New Paradigm for Data Protection' explored a new way of addressing the privacy challenges in the data world.

Rahul advises on a wide range of regulatory issues, including in relation to privacy, map regulation, fintech, encryption, spectrum regulation, e-commerce, sharing economy, biotech, digital content and streaming media. He was included in the *Mint's* list of '25 People Who Matter in Indian E-Commerce'. He is a frequent speaker and has a weekly column in *Mint* called 'Ex Machina', where he writes on technology, law and everything in between.

Rahul lives in Bangalore with his wife Ahalya, their eleven-year-old son Dhruv and their two dogs Coco and Zoey.

www.ingramcontent.com/pod-product-compliance
Ingram Content Group UK Ltd.
Pitfield, Milton Keynes, MK11 3LW, UK
UKHW050800200826
14579UKWH00010B/29/J

9 789352 779888